SOCIAL BEHAVIOUR IN ANIMALS

By
Dr. Amita Sarkar
Lecturer
Department of Zoology
Agra College
Agra (U.P.)
(India)

First Published-2003

Reprint--2004

ISBN 81-7141-747-7

Published by

DISCOVERY PUBLISHING HOUSE
4831/24, Ansari Road, Prahlad Street,
Darya Ganj, New Delhi-110002 (India)
Phone: 23279245 • Fax: 91-11-23253475
E-mail:dphtemp@indiatimes.com

Printed at:

Tarun Offset Printers,
Delhi–110 053

PREFACE

The present title Social Behaviour in Animals aims to stimulate an understanding of Behaviour of wide variety of animals including man, farm animals and best species to inspire the reader to take an interest in the field. The text integrates the descriptive and experimental approaches into a conceptual framework for the analysis of behavioural studies. It is profusely and attractively illustrated with line diagrams.

It introduces the basic ideas and concepts of modern ethology set in historical context, thereby showing how views have changed since the simple theories put forward by the founders of the field, such as Lorenz and Tinberg 40 years or more ago.

The title is not intended to be comprehensive, nor could it be at this length, but it concentrates as putting across the basic principles of the subject as briefly and lucidly as possible. It does this with the aid of carefully selected examples some recent and others classic or the field, and with numerous illustrations. The aim is to enthuse the reader with the active and exciting area of research and to lay a solid foundation on which further study of its various facts may be based.

The author has freely consulted various articles, discussion notes, reviews and extracts from the various scientific journals in the preparation of the present book, in able to make it comprehensive and upto date.

Though every care has been taken by the printer, publisher and me, it is quite likely that some errors might have found their way into the book but I hope these are of very insignificant nature. However, suggestions to improve book and pointing out of errors and mistakes, if any, will gratefully be appreciated.

The author expresses grateful to her friends and colleagues whose constant inspiration have initiated her in bringing out this book.

Special thanks are given to Mr. Wasan and staff of M/s Discovery Publishing House for their whole hearted co-operation in the publication of this book.

Author

CONTENTS

1

Group Living

The Success of Many Predators

Animal form groups to enhance their foraging success and gain between protection against predator; but group living also expose them to disease, competitions and social interference depends on surprise if the victim is alerted too soon during an attack the predators chance of success is low. This is true, for example of goshawks hunting for pigeon flocks,. The hawks are less successful in attacks on large flocks of pigeons mainly because the birds in a large flock take to the air when the hawk is still some distance away. If each pigeon in the flock occasionally looks up to scan for hawk, the bigger the flock the more likely it is that one bird will be alert when the hawk looms over the horizon. Once one pigeon takes off the others follow at once. The precise way in which vigilance changes with flock size depends on how individuals in the group spend their time. In ostrich flocks, for example, *Brain Bertram* (1980) found that each individual spends a smaller proportion of its time scanning than when alone but that the overall vigilance of the group (proportion of time with at least one bird scanning) increases slightly with group size. Therefore each bird in the flock has more time to feed and enjoys greater awareness of approaching lions (a potential predator of ostriches).

The increase in vigilance with groups size is as predicted if each bird raises its head independently of the others. The ostriches also raise their heads at random time intervals which makes it impossible for a stalking lion to predict how much time it has to creep forward undetected between look ups by its victim. Any

Fig. 1.1. The response of a flock of starlings to the approach of a bird of prey.

predictable pattern of looking could be exploited by the lion in its tactics of approach. The problem of how individuals in group scan in complicated by the fact in a large group, where overall vigilance is at the maximum value of 100 pre cent, it would pay an individual to 'cheat' loses nothing in terms of vigilance because others are busy scanning and its gains extra time to feed. It is nor known how this kind of cheating is prevented from evolving. But one suggestion is the following. Although the 'innocent' strategy of scanning regularly regardless of what others do is susceptible to cheating a flock made up of more canny individuals which do not scan unless they have seen their neighbours during the something might be resistant cheaters (*Pulliam et al.* 1982). The general point is that even when there is an overall benefit of being in a group, each individual will be expected to try to get more benefit than the others.

Dilution

With increasing group size in ostriches the vigilance increases slightly and, the chances that any one individual will be eaten during an attack by lions decreases rapidly with group size, because the lions can kill only one ostrich per successful attack. By living in a group the ostrich dilutes the impact of successful attack because there is a good chance that another bird will be the victim. To some extent this dilution effect may be offset by the increased number of attacks on larger and more conspicuous groups, but usually the net effect probably favours living in a group as the following hypothetical example illustrates. An individual antelope in herd of hundred has (all things being equal) only a one in a hundred chance of being the victim in a single attack and the herd is not likely to attack more than a hundred times as man attacks as a solitary antelope. In fact, if the herd is more vigilant it may pay the predator to concentrate its attacks on small groups and solitary individuals. The monarch butterfly (*Danaius plexippus)* migrates from North America to spend the winter in warmer places such as Mexico. They assemble into enormous communal roosts in which the trees over an area of up to 3.0 ha may be clothed in resting butterflies.

The monarch is not a very palatable butterfly, but some birds attack them in the winter roosts. Counts of the remains of predated butterflies showed that predation rate is inversely related to colony size, so the advantage of dilution seems to outweigh any disadvantage of greater conspicuousness in a large roost. The dilution effect is probably a very widespread advantage of being in a group and it might explain the strange behaviours of birds

Fig. 1.2. Struthio camelus (Ostrich).

such as ostriches and goosanders when they have young. When two females meet, each appears to try and steal the other's young and incorporate them into its own brood. Usually caring for someone else's young doesn't pay but if predation pressure is severs it might, because of dilution. A more concrete example of the dilution effect comes from a study of semi-wild horses in the Carmargue, a marshy delta in the South of France. In the summer months the horses are plagued by biting tabanid flies and during this period they are more likely to cluster together in large groups.

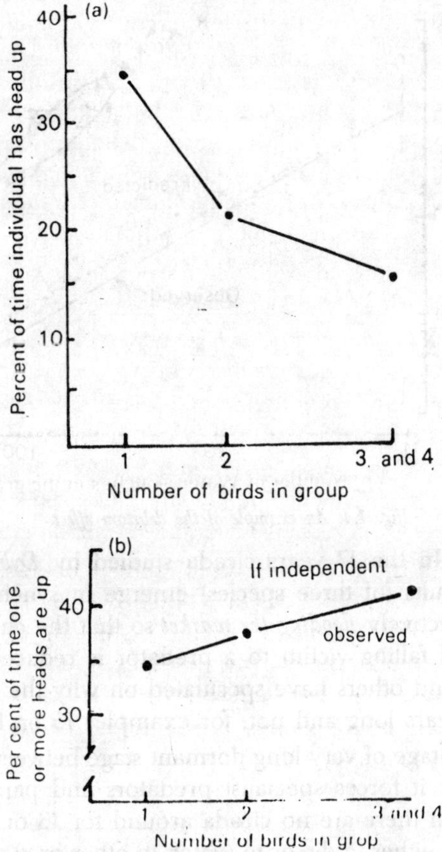

Fig. 1.3. Vigilance in groups. (a) An ostrich (Struthio camelus) spends a smaller proportion of its time scanning for predators when it is in a group. (b) The overall vigilance of the group increases slightly with group size (solid line) as predicted if each individual looks up independently of the others (broken line).

Measurements of the number of flies per horse in large and small groups showed that horse in a large group are less likely to be attacked. An experiments in which horses were transferred from large to small groups and vice versa confirmed that living in a group gives protection by the dilution effect. In certain animals dilution is achieved by synchrony in times as well as in space, and this might explain the remarkable 13 and 17 year life cycles of certain species of cicada. These insects live as nymphs underground and the adult emerge after 13 or 17 year depending on the species

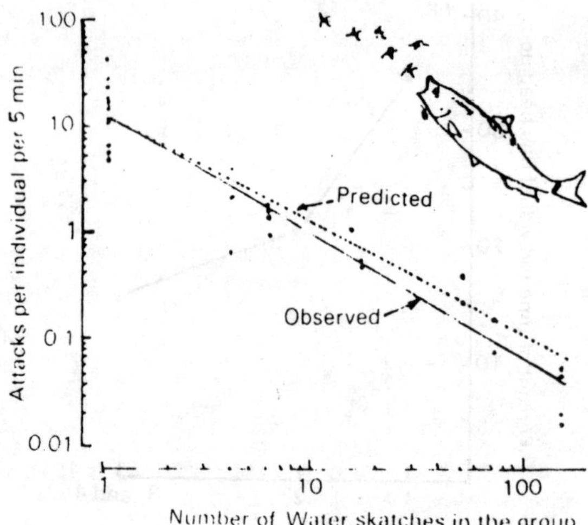

Fig. 1.4. An example of the dilution effect.

and location. In the 17 years cicada studied by *Dybas* and *Lloyd* millions of adults (of three species) emerge in synchrony over a wide area, effectively *flooding the market* so that the chances of any one individual falling victim to a predator is reduced. *Lloyd* and *Dybas* (1966) and others have speculated on why the cycle should be 13 or 17 years long and not, for example, 15 or 18.

The advantage of very long dormant stage between emergence periods is that it forces specialist predators and parasites our of business. When there are no cicada around for 13 or 17 years the predators have either to die or to switch to other prey or to become dormant themselves. The very long cycle could have evolved as a result of an *evolutionary race* In which both cicadas and their predators gradually extended their life cycles until the cicadas eventually won. The significance of the 13 and 17 year periods is that these are prime numbers which means that a predator could not regularly fall into synchrony with the cicadas if it had a short life cycle of which the cicada cycle is a multiple if for example cicadas had a 15 years cycle, predators with 3 or 5 year life cycles would fall into step with their prey every fifth of third generation. This idea remains an interesting speculation but synchrony is certainly an advantage. Field evidence shows that cicadas emerging at the peak of the clue have a lower chance of succumbing to predators than those emerging early or late. Section therefore acts to maintain synchrony once it is established. Just as a cicada in the middle of the emergence period is safer than one at either end, individuals in the middle of a flock, school or herd may enjoy greater security than those at the edge.

If the predators pick off victim from the edge, each member of the group should jockey for accentual position and, in effect seek cover behind the other (*Hamilton* 1971). This may explain why starling flocks for example bunch together in a tight group when a predator approaches. Why should predators attack the edge of the group? The old trick to throwing three tennis balls to a friend at the same time shows how difficult it is to track one of a number of rapidly moving objects in the visual filed for long enough to catch one. There is some evidence that predators suffer from the same type of confusion when attacking a dense group of prey and this may provide an explanation of why attacks should be directed at the edge of a group.

Group Defence

Prey animals are often not just passive victims and by living in a group they any be able to depend themselves against the unwelcome attentions of a predator. In colonies of blackheaded gulls nesting Paris will mob crows when it flies near their nest and in the center of a dense colony many gulls mob the crow at the same time because it is close to many nests. The effect of this is to reduce the success of the crows in hunting for gulls eggs.

Fig. 1.5. Conis lupus (Wolf).

Costs of Being in a Group

Social behaviours evolves when individuals that join with others survive and reproduce better than those that do not. Determining the benefits and costs of group living is not easy because same behaviour may be advantageous for individuals as one species but disadvantageous for those of another species or even for the same species at a different time. This cost of being in a group was studied experimentally by *Malte Anderson* using artificial nests of the fieldfare, a thrush like bird which breeds colonially in Scandinavian boreal forests. The bulky nests are quite conspicuous and a colony of artificial nests attracted more predators than did solitary nests. However, fieldfares vigorously mob and defecate on crows and other predators and *Anderson* and *Wicklund* found that artificial nests placed near a colony of fieldfares survived better than those placed near solitary fieldfare nests. They concluded therefore that the benefit of group moving by members of a colony more than offsets the disadvantage of being conspicuous. This is supported by *Volker Haas's* (1985) observation that nesting success is higher for colonial than for solitary fieldfares.

Fig. 1.6. Canis dingo (Dingo).

Living in Groups and Getting Food

Findings Good Sites

Many species which feed on large ephemeral clumps of food such as seeds or fruits often live in groups. For these animals, the limiting stage in feeding is the problem of finding a groups. For these animals, the limiting stage in feeding is the problem of finding a good site: once the patch has been found there is usually plenty of food, at least for a short while. *Peter Ward* and *Amotz Zahavi* developed the idea that communal roosts and nesting colonies of birds may act as '*information centres*' in which individuals find out about the location of good feeding sites by following others. The idea is that unsuccessful birds return to the colony or roosts and wait for the chance to follow others who have had more success on their last feeding trip. Unsuccessful birds might recognize successful ones by, for example, the speed with which they fly out from the colony on their next trip. *Ward* and *Zahavi* used the phrase '*information centre*' which carries with it a connotation of mutual co-operation in the transfer of information, as for example in a honeybee or ant colony. There are special reasons to expect co-operation in social hymenopteran colonies.

'*Mutual parasitism*' might be a appropriate label here, since the successful foragers are in effect parasitized by unsuccessful birds. Each individual is out to maximize it own success and not the success of the colony as a whole. In some species the *informer*' might be unable to avoid being followed because it is conspicuous when it leaves the nest, for example seabirds leaving a colony on a cliff. The informer might, however, benefit from being followed. The benefit could be long term: on later trip the leader becomes follower, or short term: there may be an advantage to feeding in a group because, for example, of the reduced risk or predation. If these benefits do not outweigh the disadvantage of the competition at the feeding grounds which arises from being followed, a successful bird should conceal as much as possible information about its success. *Ward* and *Zahavi's* worked communally roosting weaver birds (*Quelea quelea)* by *Peter de Groot* (1980).

Quelea nest in colony and roost in groups sometimes estimated to contain over a million birds. They are serious agricultural pests and can devastate a grain filed in a few hours De Groot's experiments were done on a some what more modest scale. Two groups of birds roosted together in the large aviary labelled X and

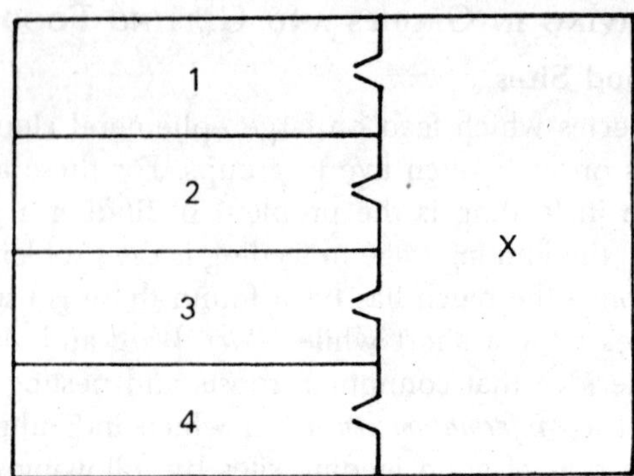

Fig. 1.7. An experiment to test the 'information centre' hypothesis with Quelea. The birds roost in the large area labelled X and feed in the smaller compartments labelled 1-4.

had access to foraging areas in the small compartments labelled 1-4. The birds could not see into the foraging compartments from the roosting area but had to pass through small entrance funnels to explore them for food or water. In one experiment one of the groups (A) was trained to find water in one of the four compartments and the other group (B) was trained separately to find food in mother compartment. The two groups were allowed to roost together and were allowed to roost together and were deprived of food or water. When they were thirsty, the birds of group B followed A to the drinking site, and when they were hungary A followed B to the feeding site. Somehow the 'native' birds assessed that the other group was knowledgeable and followed it to the resource supply. In a second experiment group A was trained to forage on a good food supply (pure sed) in one of the compartments while group B was separately trained to fly to another compartment for a poor food supply (seed in a bed of sand). When the two group followed the first group when they left the roost at dawn. It is nor yet known how the birds recognize which individuals to follow. The mode of information transfer has been identified in another study.

Geoff Galef and *Stephen Wigmore* (1983) trained rats *(Rattus norvegicus)* to search for food in a three-arm maze. Each arm had food with a different flavour, cocoa in one, cinnamon in another, and cheese in the third. In the first part of the experiment the rats learned that on any particular day only one of the three sites contained food, but the site was unpredictable. Then on the days of the actual experiment each of the seven test rats was allowed to sniff a *demonstrator* rat in a neighbouring cage. The demonstrator had been allowed to feed on whatever randomly chosen food was available for that day, and four of the seven test rats, having sniffed demonstrator, went to the correct site on their first choice of the day. *Sniff* is the operative word, because other experiments showed that the cue the test rat picks up from the demonstrator is the smell of the food it has eaten, just as you can tell when your friend has had a garlic pizza. Learning about potential sources in a more direct way, by seeing others exploiting them, is important in both flocks of birds and shoals of fishes. Individuals in groups may be able to capture prey which are difficult for a single individual to overcome either because the prey is too large for one predator to handle (e.g. Lions hunting adult buffalo) or because it is too elusive for one predator to catch (e.g., killer whales hunting porpoise). When the prey themselves are in a group the predators may, by hunting in a group succeed in separating a victim from its companions and subsequently chasing it until they overtake it. This is how predatory fish such as the jack (*Caranx ignobilis*) hunt for schooling prey. Individuals in a group are more successful than single fish when hunting for schools for Hawaiian anchovy (*Stolephorus purpureus*). However, the benefit is not shared equally between members of the hunting group: fish at the front of the school when it is chasing a prey get more than those at the back. In fact, the fourth and fifth fish could do better by hunting alone, but it may be that different individuals occupy the lead position during different chases. This is a reminder of the general point that benefits of being in a group may not be shared equally.

Harvesting Renewing Food

There is a variety of food which renews its self continuously after being eaten up by the animals for example growing vegetation. The amount of food available in a site increases with time since the last visit, so an individuals could get the maximum possible foraging returns by coming back to the same site after the appropriate time interval. Returning too soon means not finding enough food and returning late means missed opportunities to eat a plentiful food supply. As we saw with the wagtails in chapter 5, the problem wit harvasting a renewing food supply in this way is that it only works if there is no interference by others with the renewal pattern, individual A's strategy of returning after ten days would fail if B visited the site after, for example nine or eight days.

One way to prevent interference by others is to defend a territory and another way is to visit sites in a group so that everyone returns at the same time. Wintering flocks a Brent geese (*Branta bernicla*) feeding on salt marshes in Holland seem to do the later, where territory defence would not be feasible because the marsh is frequently inundated at high tide. Continuous observation of 40 one hectare plots from dawn to dusk for 24 days during the spring showed that the flocks return to exactly the same site on the marsh at regular 4 day intervals. No only does this allow the sea plantain (*Plantago maritima)* to recover between visits by the geese, but also the regular cropping pattern actually stimulates the growth of young leaves which are rich in nitrogen.

Experiments in which sea plantain was cut with scissors to simulate goose grazing at different time intervals suggested that given the average bite size, the geese may even return after a time interval which maximizes the growth of young shoots. As stated already that in the hunting school of fishes the individuals at the back of the group fare less well than those at the front, this feeding behaviour is seen in others animals too. Now it has been well probed that the overall benefit in terms of food intake is similar for birds in different parts of the flock. The front birds eat most of the vegetation, but the youngest and most nutritious parts of the *Plantago* plants are the base of the leaves close to the ground. These parts are exposed only after the older, taller, parts of the leaves have been cropped off, so it is reasonable to hypothesize that the first birds eat larger mouthfuls while the later birds eat food of higher quality. The overall effect may be that all birds obtain the same quantity of nutrients.

Costs Associated with Feeding

The potential cost of feeding in a group has been suggested in goose as far as the competition for food is concerned. Competition may take the form of direct exploitation as in the jack where fishes at the front catch the prey and deprive those at the back of the school or it may arise as a result interference in which the availability of food to a group member is reduced as a result of the behaviours of nearby companions. This occurs in the redshank (*Tringa totanus),* a shorebird studied by *John Goss-Gustard.* Redshank feed in tight flocks at night and more loosely scattered or solitarily

Table 1.1. Example of studies of students in which possible costs and Benefits of Group Living other than those mentioned in the text have been measured.

Hypothesis	*Test*	*Reference*
1. Warm booded animals save energy because of thermal advantage of being close together	Pallid bats (Antrozous pallidus) roosing in groups use less energy than solitary roosters	Trune & Sloboduchikoff (1976)
2. Inferior Competitor can overcome competitive advantage of another species by group foraging	Blue tang surgeon Fish (*Acanthurus caeruleus*) are excluded from algal mats by dusky damsel fish (*Stegastes dorsopunicans*) when solitary but not when in a group	Foster (1985)
3. Hydrodynamic advantage for fish swimming in a school. They save energy by positioning themselves to take advantage of vortices created by others in the group	Measurements of distances and angles between individuals show that they are positioned to benefit not correctly	Weihs (1973) Patridge and Pitcher (1979)
4. Increased incidence of disease as a result of close proximity of others	Measure number of ectoparasites in borrows a prairie dogs (*Cynomys* spp.). There are more parasites per burrow in larger colonies	Hoogland (1979b)
5. Risk of cuckoldry by neighbours	In colonial nesting red-winged *blackbird* (*Agelaius Phoeniceus*) the mates of vasectomized males laid fertile eggs. They must have been fertilized by males other than their mates.	Bray et. al. (1975)
6. Risk of predation on young by cannibalistic neighbours	In colonies of Belding's ground squirrels (*Spermophilus belding*) females with territories are move likely to lose their young to cannibalistic neighbours than are females with large territories around their burrows.	Sherman (1981)

during the day: this difference seems to be related to interference. In the daytime, redshank feed by sight on small shrimps (*Corophium*) which live with their rails just sticking our of the mud surface, while at night, when visual search is impossible, the birds turn to feeding by touch on snails (*Hydrobia),* sweeping their long beaks through the mud. When feeding on shrimps, the birds are likely to interfere with one another because the shrimps retreat into the mid an become unavailable as soon as they detect the heavy clump of redshank feet. Feeing rate is therefore greater with increasing neighbour distance and the birds tend to space out. At night there is no feeding interference because the birds do not depend on seeing the prey, and the snails in any case do nor react quickly to disturbance.

Feeding rate in these conditions is not related to flock density and the birds crowd into tight groups One interpretation of these results is that there is an advantage to tight clumping in redshank and that during the night there is little cost to feeding close together causes interference and the birds spread out. The nearest neighbour distance under these conditions appears to reflect a balance between the costs and benefits of group leaving. It has also been studied that the competition for food influences group as seen by *Mark Elgar* (1986) in house sparrow (*Passer domesticus)* flocks. The first sparrow to arrive at a feeding site on the ground gives a *'chirrup'* call from a perch before going down to feed. The chirrup draw in other sparrows and the group goes to the food source together. It may seem maladaptive for a sparrow, having found a good source of food, to attack other competitors to it, but it is known that sparrows spend less time scanning in larger groups so there is some benefit to the first sparrow in generating a flock in which to feed. However, and this is the most interesting point, the first sparrow does not give a chirrup to recruit others if the food source is indivisible. The cost of recruiting in terms of competition for food now outweighs the benefits derived from reduced scanning time. Elgar's experiment was simply to provide the same amount of food-and piece of bread-either in a lump (indivisible) or in crumbs (divisible). The first sparrow to arrive was more likely to chirup when the food was in crumbs.

Optimal Groups Size

There are many different costs and benefits of living in a group, some or all of which might be relevant to a particular species.

Pigeons, horse ostriches and cicadas do not necessarily get together for the same reasons, but they could do so for any or all of the reasons discussed. Our list has been by no means comprehensive; we could have gone on to describe costs of group living such as transmission of diseases, cannibalism, and cuckoldry, or benefits such as protection from the elements, co-operative territorial defence, and increased efficiency of locomotion. But rather than try to extend the list of possible costs and benefits to tis exhaustive and exhausting conclusion, we will return to the more interesting question of wether different kinds of cost and benefit can be combined to predict a optimal group size.

Comparative Studies

The costs and benefits of living in a group interest and a qualities picture can be drawn from comparisons between species. For example some species of shorebirds such as the knot feed in dense flocks while others such as the ringed plover feed in loose flocks or as solitary birds. It is known that living in a flock confers protection on shorebirds against attacks by birds of prey so why do not all species feed in tight flocks? The species which feed in dense flocks hunt by touch and walk slowly, probing or swinging their beaks through the mud, while solitary and loose flock feeders hunt by sight and move rapidly, picking prey off the surface of the mud or water, perhaps, as with the redshank, the costs of feeding interference are so big in the latter species, that the net benefit for an individual is higher when alone, even though the risk of predation is greater.

Time Budgets

Ron Pulliam (1976) and *Tom Caraco* (1979a) have developed a model of optimal group size base on time budgets. They used time as a currency. The model is meant to illustrate the factors influencing winter flocks of small birds. The survival of birds in a flock is considered to be dependent on two main risks, starvation and predation and the birds' time budget is divided into three types of behaviour associated with these risks: scanning (for predators), feeding and fighting (for food). Based on their observations of yellow-eyed junco flocks, *Pulliam* and *Caraco* divide fighting into two categories: short term squabbles over attempt to evict subordinates from good feeding sites in order to ensure a supply of food for the rest of the winter. The three activities in the time budget are assumed to be mutually exclusive, that is a bird

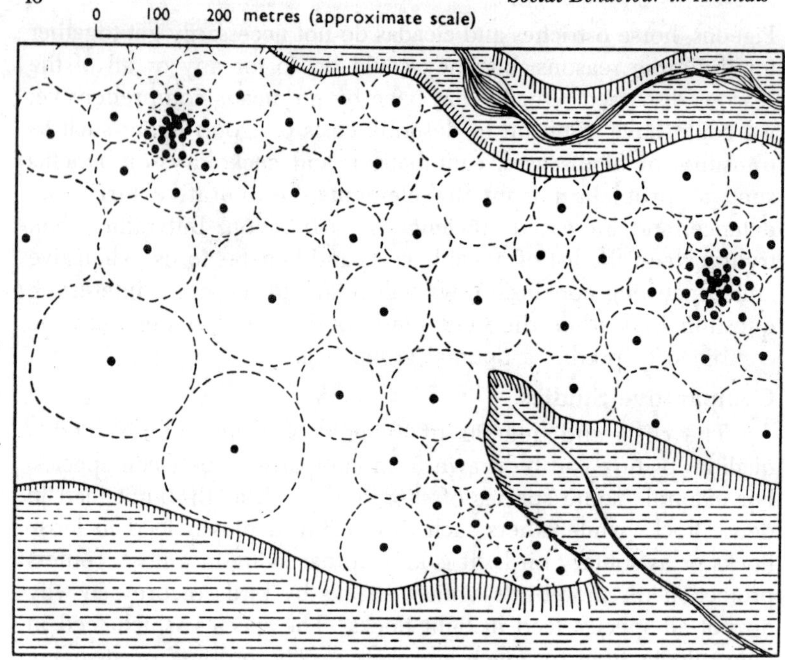

Fig. 1.8. Territorial map of males of the Uganda kob on an area of grassland raised above two swampy river bed areas, indicated by shading.

cannot, for example, *acan* and *feed* at the same time. In order to scan it has to point its head upwards, while pecking involves facing towards the ground.

Finally they assume that scanning for predators takes precedence over feeding since failing to see an approaching predators is more dangerous than failing to eat a seed. Dominant birds are assumed to give higher priority to satisfying their daily energy requirements than to long term eviction subordinates, while for a subordinate bird, aggression must take priority over feeding since a bird cannot feed while it is being attacked. The main features of the model are as follows:

1. The proportion of time spent scanning by an individual is assumed to decrease with increasing group size. The basis for this assumption is that a given level of vigilance can be maintained with less scanning per individual as group size increases.
2. As group size increases and encounters between birds become more frequent, the proportion of time spent in aggression increases.

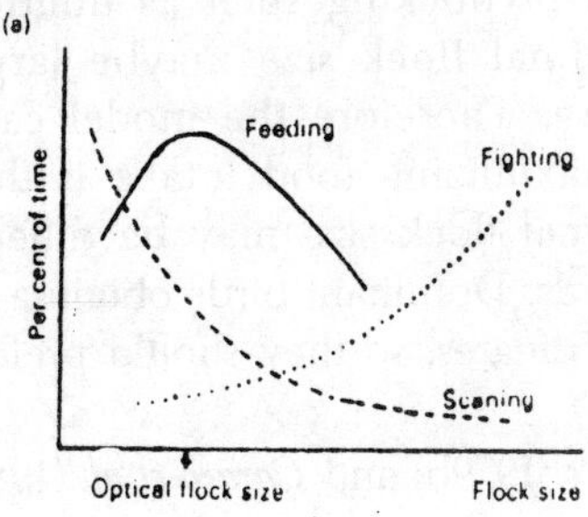

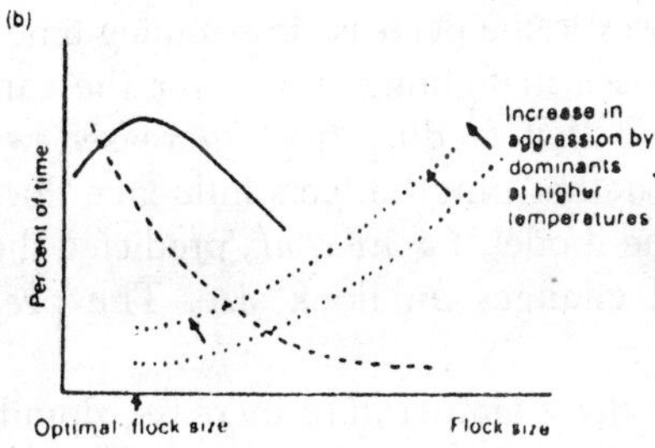

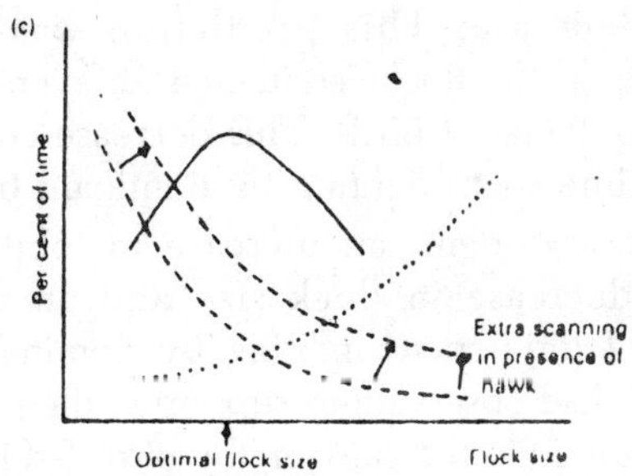

Fig. 1.9. A model of optimal flock size (a) As flock size increases birds spend more time fighting and less time scanning. An intermediate flock size gives the maximum proportion of time feeding. (b) At higher temperatures dominant birds can afford to spend more time attacking subordinates. The optimal flock size for the average bird therefore decreases. (c) When predation risk is increased by flying a hawk over the flock, the scanning level should go up and the optimal flock size is increased.

3. The time spent feeding therefore is at a maximum in flocks of intermediate size.

If the benefit of flock feeding is to increase the time available for feeding while maintaining a certain level of vigilance. If there are other benefits of flocking such as dilution and increased vigilance, the optimal flock size maybe larger than the one illustrated in Figure. Therefore the model can be used to test whether or nor maximizing food intake is the only benefit of flocking. The optimal flock size may be different for dominant and subordinate birds. Dominant birds obtain a long term benefits from evicting subordinates, so they should prefer to be in smaller groups.

However, *Caraco* (1979b) and *Caraco et al.* (1980a) have recorded the time budgets of yellow eyed juncos in winter flocks. They found that the proportion of time spent by individuals in scanning and fighting change with flock size in the directions assumed by the model. However, the decrease in scanning time was much greater than the increases in fighting time over the range of flock sizes they studied, so that feeding time increased with flock size. In order to test whether time budgets influence flock size in the way suggested by the model, *Caraco et al.*, predicted the effect of various environmental changes on flock size. The predictions were as follows:

1. As average daily temperature increase, dominant birds should have more time to evict subordinates because they can satisfy their energy requirements more rapidly. Flock size should therefore decrease. This prediction was supported by observations; at 2°C flocks contained an average of 7 birds at 10°C they contained 2 birds. This decrease coincided with an increase in time spent fighting by dominant birds.
2. By a similar argument, an increase in food supply should produce a decrease in flock size and an increase in the proportion of time spent fighting by dominant birds. Again the results of field observations supported the prediction. When food was scattered in the canyon, the birds fed in smaller flocks.
3. An increase in the risk of predation should have exactly the opposite effect of the previous two changes. This is because a high risk of predator attack should cause the birds to spend more time scanning; they therefore have to feed in larger flocks to maintain a given rate of food intake *Caraco at al.* (1980a) allowed a tame hawk to fly over the canyon, and as predicted the birds spent more time scanning and the mean flock size increased. It was 3.9 birds without the hawk and 7.3 with the hawk.

4. Finally, *Caraco* predicted that by adding more cover to the canyon in the form of a bush, the effective risk of predator attack would be reduced because the juncos would have easier access to a safe hiding spot. The birds should therefore spend less time scanning which allows more time for feeding and fighting. This is exactly what happened when an experimental bush was placed near one of the favoured feeding sites, and as expected the flock size decreased.

From these results we can conclude the influence of time budgets on the flock size as hypothesized in the model. Flocking allows more time for feeding because less time is spent scanning and the maximum flock size depends on the time available for dominant birds to evict subordinates. Secondly, the results allow us to reject the simples hypothesis about optimal flock size. The birds did not feed in flocks of the size that would have maximized feeding time: under normal conditions the average flock in canyon contained 3.9 birds, but measurements showed that the time available for feeding would have been higher in a flock of 6 or 7. As we have mentioned already, the optimal flock size for dominant and subordinate birds probably differs since dominant birds benefit by evicting subordinates. The observed flocks may have been a compromise between the optimum for dominant and subordinate birds.

A further complication is that birds in larger flocks benefit by the dilution effect and increased vigilance, as described earlier for ostriches. On the basis of model stated already we see that time budgets can be used to analyse the effects of different costs and benefits on flock size. I also reminds us of the idea that flocking and resource defence may be two ends of a continuum. The model could be viewed as one which predicts the conditions under which it pays dominant birds to exclude subordinates and defend a territory. When food is plentiful or predation risk is low, dominants can afford the time to maintain a defended area or in other words the territory is economically defendable. *Richard Sibly* (1983) has pointed out that the groups of optimal size may rarely be found in the wild because if there was a group of this size, it would pay solitary individuals around to join the group and therefore push the group above the optimal size. The curve shows a possible relationship between individual fitness (perhaps measured as feeding rate of probability of escaping predation) as a function of group

Fig. 1.10. Some mixed species flocks involve complex relationship.

size. The optimal group size, in which average fitness is maximal, is seven individuals. Imagine that individuals are free to join the group or forage alone as they arrive in the area.

Clearly, if each individual chooses the option that maximizes it fitness, newcomers will join the group up to the point where group size is 14, twice the optimum! At this point the pay-off for foraging alone is the same as the pay-off joining the group. The same principle would apply if one imagined groups splitting and reforming as smaller units, but the argument is more complicated. A group of 12, for example would split into two units of 6, but then one individual would be joined by another because 8 is better than 5, and so on. The exact outcome depends on the shape of the curve and it is possible to draw fitness curves which will result in a stable group size (*Giraldeau* and *Gillis* 1985). The general point, however, is that one should expect to find stable groups rather than optimal groups in nature and very often, observed groups will be larger than the optimum.

Individual Differences in Group

It has already been told that individuals in a group may benefit from group living to different extends. In foraging groups of jack, individuals at the front do between than those at the back. Similarly in starling flocks, birds at the edge spend more time scanning than those in the middle (*Jennings* & *Evans* 1980). How are such differences between individuals maintained? In many cases, they are simply a reflection of dominance relationships. Within the group, older more experienced, or bigger individuals are able to commandeer the best positions and force others to take what is left. Subordinate individuals put up with less pay-off as long as they could not do better by moving elsewhere. An alternatives view of individual differences is that different individuals achieve similar pay-offs but in different ways. In sparrow flocks, for example, some birds are good at locating new sources of food while others are good at stealing food once it is found. These two strategies, producer and scrounger, may coexist in the flock with equal pay offs for the two.

2

FIRST VOICE

About 350 million years ago, in the steaming primaeval swamps, a new group of vertebrate animals was beginning to evolve the abilities which would enable them to live on land–they were to become the amphibians. At that time, fish, crabs, lobsters and shrimps living below the water, and insects and spiders already on land, were making their instrumental scrapings and raspings. But, some 200 million years ago, the first voice in the history of the earth was probably that of an amphibian–may be something liked that of a frog. Frogs, it has been said, divide their world into three classes of objects–if it's small enough you eat it; if it's too big you run away from it; and if it's intermediate you mate with it. In order that frogs mate with the right object, namely a female of the same species, they rely on sound. There are frogs and toads that chirp, others croak; some whistle; and there are those that click.

Some species produce constant pure tones; others very noisy broad-band calls. Two species of *Kassina* frogs from South Africa, the predominantly terrestrial *K. senegalensis* and the aquatic *K. maculata* have frequency modulated call *K. maculata* emits a mating call that sweeps upward in frequency from about 800 Hz to over 3,000 Hz in approximately 15 milliseconds, a feat to rival sound production in birds and bats. Indeed, the sweep is very much like the modulated sweep of a bat except that it lies in the range of audible frequencies.

The Auditory System of Frogs and Toads

Frogs make their sounds by forcing air from the lungs across vocal chords, into the buccal cavity. By keeping the mouth closed and shutting the external nares the air can be shunted back and forth across the vocal chords to produce the characteristic chirping sounds. These can be further enhanced with the aid of a single vocal sac below the chin or a pair of balloon-like sacs on either side of the mouth, which serve to amplify the sound. A natteriack toad *Bufo calamita*, without a vocal sac, reaches to only 150 metres.

Frogs croak to attract females

To avoid being spotted by predators, most frogs and toads mate at night when it would be ineffective to lead around or perform some visual display. So the best way to attract attention is by a sound signal. One of the pioneers of research into the calls of frogs and toads is Dr. Bob Capranica from Cornell University, New York. His aim has been to learn something about the role that sounds play in the lives of frogs and toads and then to discover how they locate and recognise those sounds. Capranica started at the University of California, Berkeley, graduating in electrical engineering. He became interested in communication theory and went to Bell Telephone Laboratories, designing electronic communication systems. With a Bell Labs Fellowship he was able to go to the Massachusetts Institute of Technology where he met biologists and medical researchers interested in speech. He decided to pursue studies of communication in animals. MIT's novel analytical tools were to change the nature of this field of research. No-one at the time was approaching studies of animal communication at the level of sophistication at which scientists were pursuing, for example, studies of speech.

Ken Stevens had built the first computer to synthesis speech sounds, and Capranica thought that these techniques might be useful in studying animal sounds. His first study animal was to bullfrog *Rana catesbeiana*, the largest of the North America frogs. He built indoor terraria, and brought frogs into the laboratory. At first, Capranica noticed that occasionally a male bullfrog would call with a deep, resonant croak and another male in the laboratory would be triggered to call. Capranica called this the 'evoked calling response'. He carried out a series of simple playback experiments, exposing male frogs to the recorded call of others. The males always called back. They wouldn't response'. He carried out a series of simple playback experiments, exposing male frogs to the recorded call of others. The males always called back. They wouldn't respond, though, to the recorded calls of other species of frogs or toads. The vocalisations of bullfrogs, at least, Capranica concluded, are

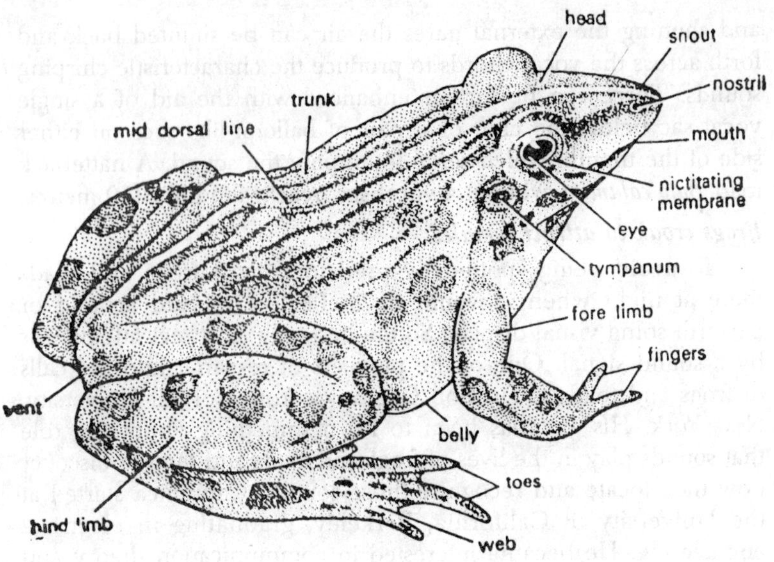

Fig. 2.1. An Indian frog, Rana tigrina.

species specific. There is also the suggestion that male bullfrogs recognise individuals by differences in their calls. Males are territorial and fight, initially with neighbours, to establish their boundaries around the choice territories.

Bullfrogs prefer to spawn in areas of the pond that are least subject to extreme variations in temperature, which might cause abnormalities to appear in the developing embryos, and that are least likely to harbour embryo-eating predatory leeches. Females appear to choose males on the basis of the quality of the territory they have acquired. Older and larger frogs muscle-in on the best territories. After a period of vocal contest with neighbours male bullfrogs tend to get used to each other's calls and only act aggressively towards intruding strangers. If a neighbour's call is played to a resident male, generally it will be ignored, the resident continuing to call to attract a mate. If a stranger's call is played the resident will stop singing and rush towards and threaten the loudspeaker. These calls, however, are not the only sounds emitted by bullfrogs. Capranica identified other different vocal signals such as an alarm call, a distress call, associated with feeding behaviour, and the release call. The distress call is made by a bullfrog when it is a very high-pitched scream, lasting several seconds, a little like a baby's cry.

The scream is a startling sound, the only call made with the mouth wide open. If a bullfrog is grabbed by a racoon, or some other predator, the scream can be heard across the pond. Whether this is to warn others of danger or an attempt to startle the predator is not clear. The other interesting sound made by bullfrogs is the release call. Male bullfrogs are not very good at establishing the sex of a silent frog they may encounter on a dark night. If another frog should stray into the male's immediate vicinity and touch him, the calling male will stop calling and clasp the intruder in the hope that it is a female. If it is the right sex the female remains silent and the male retains his hold. If, however, it is an unreceptive female, may be one already having spawned, or a male, then the animal will produce a distinct release call. This lasts for several seconds. The male recognized•an inappropriate partnership and the animal is released. At the American Museum of Natural History, New York, researchers carried out a rather entertaining experiment which verified the significant of the release call. They took several balloons and partly filled them with air and water so they would float just below the water surface. A string was tied at one end and the balloon dragged along near to a calling male frog. The male stopped his calling and clasped the balloon. Since it couldn't produce a release call, the male sat there holding the balloon for the entire evening! The experiment clearly showed that the release call is necessary for discrimination of an appropriate receptive partner.

Similar observations have been made in Britain with common toads. They will hold on tightly to any small moving object that does not emit a release call. They have been seen to attempt mating with goldfish and even Wellington boots! Capranica recorded and analysed all the bullfrog calls and then, using the same techniques as the speech researchers, proceeded to syntghesise them. Thus he was able to vary different parameters in order to find out which components of the sounds the bullfrog recognised as characteristic of its own species. On returning to Bell Telephone Laboratories, after completing his doctorate at MIT, Capranica met a researcher from the university of Haifa, Israel, Eviatar Nevo. Nevo claimed that he had detected, simply by ear, geographic variations in the mating calls of cricket frogs *Acris* spp.

Nevo and Capranica spent three years establishing that such variations existed. They made over 5,000 recordings and took them to Bell Labs for analysis. Then they started a series of playback

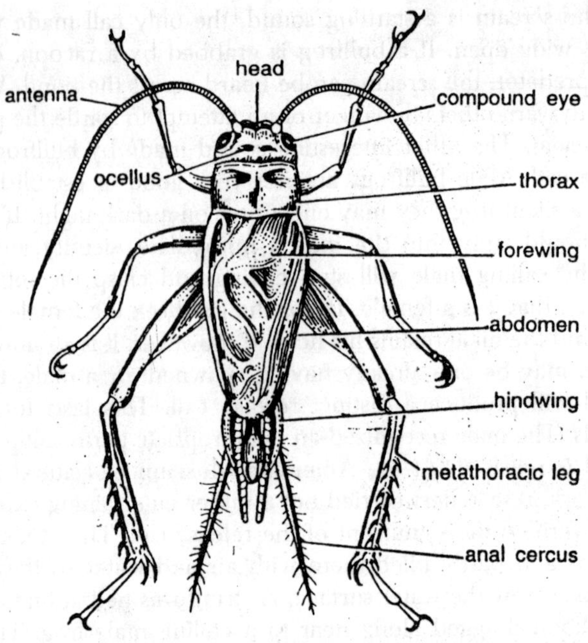

Fig. 2.2. Gryllus. (Dorsal view).

experiments. Two species of cricket frog occur in the USA, *A. crepitans* is found throughout most of the USA east of the Rockies, and *A. gryllus* is located mainly in the south-east. During the spring, cricket frogs emerge from hibernation and congregate at their breeding ponds. The males begin to call. Populations of frogs, with many species present, can sometimes be quite dense so each male must compete with many others for his signal to be audible. The female enters the pond and selects a calling male of her own species. She may sit just six to eight inches away from him, listening to his repetitive calls, sometimes staying for an hour or two. Finally, when she has made up her mind that this is the frog for her, she swims the last few inches making physical contact with the male. He immediately stops calling and clasps the female in amplexus for up to three hours.

As is the usual pattern in frogs, the female spawns and the male fertilizes her eggs externally. All this happens on one night of the year only. The female is only responsive to a male's call on that one night. Therefore, in order to test the female's discriminating abilities, Nevo and Capranica searched ponds and lakes for pairs of cricket frogs in amplexus. Clearly, the female is likely to respond more readily to playbacks at this time as her ovulatory cycle is in full swing, and she will spawn in two or three hours. If separated from the male at this point there is tremendous pressure for her to find a mate immediately. Separated females were brought to an arena away from the calling males in the pound and placed between two loudspeakers on the ground. Through one loudspeaker was played the mating call of a male from the female's own area, and through the other the call of a male from a different area.

The female was released and hopped immediately towards, one of the speakers–sometimes scratching at the loudspeaker in order to get at the male presumably calling from inside. Using both natural and synthesized sounds, Nevo and Capranica showed that female cricket frogs would respond to the calls of males of their own local population rather than those of males of different geographical populations. Females in central Texas, for example, would react to the calls of males from central Texas, but not to the males of east or west Texas, or Alabama, or Louisiana. Cricket frogs, it seems have geographical dialects. By analysing the calls, the researchers identified a stereotyped pattern of clicks which is distinct, and recolgnised by the female, from each dialect area. Nevo and Capranica used sound to compare the evolution of both species and looked at how the characteristic of the sound vary geographically in different directions. In this way they could attempt to pinpoint the original location of each species. After analysis of 5,000 calls, one of the most complicated analyses of animals sounds so far undertaken, it turned out that cricket frogs started out in a region around Georgia, in a south-eastern part of the USA and radiated from this local focus. *A. crepitans* spread throughout most of the USA, whereas *A. gryllus* moved down through Florida. Furthermore, the analysis included such environmental factors as climate, humidity, average rainfall, temperature etc. and it emerged that geographical variation in the calls can be linked with environmental factors the sounds that each local population makes are adapted to the environment in which the animals live.

Since certain frequencies or patterns of calls travel better under particular circumstances, the animals have evolved calls that maximise of long-distance communication. Human ears are fat apart, and a large solid object between our ears, the head, helps to provide a sound shadow. We assess the differences in intensity between the two ears, and identify minute differences in the arrival of sound waves at each ear. But what about frogs? How does such a small creature localise sounds? This was a puzzle as the cricket frogs are about two centimetres in length from snout to vent, weight just one gram, and the distance between their ears is less than a centimetre. In an attempt to find an answer, Capranica turned his attention to the mating calls of another North American frog. Living alongside the cricket frogs in Georgia, Alabama, and Florida is the green tree frog *Hyla cinarea.* Together with Carl Gerhardt, of the University of Missouri, and Jurgen Rheinlaender, or Ruhr University, West Germany, Capranica attempted to identify the features of the call that a female recognised, how well she localise the call, and what it is in the calls that permits localisation. Again, pairs of green tree frogs in amplexus were found.

The female was separated, and mating calls played back to her through a movable loudspeaker with a video camera, and under very dim light. The arena was criss-crossed with grid-marks so that her trajectory could be traced very accurately. To their amazement, the research team found that a female green tree frog would orientate and hop towards the loudspeaker with an accuracy of about 10, that is, she could pinpoint the source of a male's call to within an arc of less than 10(, and from four metres away. For such a small creature to locate with this accuracy is quite remarkable. Humans can localise sounds directly in front with an accuracy of about 3(, and sounds to the side of about 7. The green tree frog is a tiny little animal with ear drums exposed directly to the air, flush with the head (an adaptation for streamlined swimming), and very close together. In addition, the sounds being used are relatively low frequency sounds with long wavelengths. The researches continued their experiments using a female which had already proved her skills by locating the loudspeaker to within 10°. She was re-released into the arena, only this time with grease placed on one eardrum. This, in effect, temporarily deafened that ear. The female became disorientated and hopped around in circles. If the left ear was covered by grease she would circle to the right; if the grease was on the right ear she would circle to the left. As she circle, she would continually wipe the side of her head with each leg in an effort to dislodge whatever was causing her hearing impairment. When the grease was removed and the female placed yet again in the grease was removed and the female placed yet again in the arena, she would once more go directly towards the sound.

Rheinlaender, Gerhardt and Capranica had verified, for the first time, that small vertebrate animals, like frogs and toads, require both ears and must carry out some binaural comparison somewhere in the nervous system; all the more remarkable, for the time of arrival differences at the two ears separated by less than a centimetre is only of the order of one or two microseconds, and the intensity difference a very small fraction of a decibel. It is hard to understand how the nervous system of such a simple creature could process such incredibly small cues, yet it appears that frogs can do it. How, in detail, they do it remains unsolved. It has been suggested that there is a pressure gradient system operating. Sound strikes one eardrum causing it to vibrate. Some sound goes through the eardrum, out the other side, along the Eustachian tube, and vibrates the other eardrum from the inside. Sound from both inside and outside reaches each eardrum. The sound on the outside compared with the sound on the inside has a different phase relationship because of the additional path length that the sound has to travel inside the tubes.

As a sound source is moved around the frog in the external world, so too does the phase difference across each eardrum change. In additional, some sound is thought to enter through the nose, and travel via the mouth cavity, along the Eustachian tubes to strike the inside of the eardrum. So sound is coupled, not only from one eardrum to the other, but also through the nose. By using this clever technique small animals are able to achieve the same directional sensitivity as larger animals, without the external pinnae and a large head. More recent work by Gerhardt and Rheinlaender has demonstrated that female green tree frogs can also localise elevated sounds by scanning with the head, sometimes with the jaw parallel with the ground and sometimes with the head lifted. The eardrum itself, though, is very large indeed. A typical male bullfrog, for example, has for its size one of the largest eardrums of all vertebrates with a diameter of over two centimetres. The female's is about half that size. This sexual dimorphism is only present in a few species; more normally the two sexes have the same size eardrums. By bouncing a laser beam off the male bullfrog's eardrum, Capranica and his colleagues have shown that the amplitude of vibration is also very large. A sound of 90 dB causes the membrane to vibrate with an amplitude of 15,000 Angstroms. This is so large a movement that it can be observed through a light microscope.

For comparison, a cat's eardrum, exposed to the same intensity of sound, vibrates with an amplitude of only 400 Angstroms. Why a bullfrog should have large eardrums that vibrate so violently is not clear. The human ear has a threshold sensitivity which is more or less at the optimal theoretical limit. If it was more sensitive we would hear the molecules randomly striking the eardrum. Our auditory threshold sensitivity is, by definition, zero decibels. If our ears were ten decibels more sensitive, this background noise would interfere with communication and our ability to detect sounds at lower frequencies. Bullfrogs, despite their enormous eardrums vibrating at large amplitudes, have a low sensitivity of 30 decibels. Other frogs and toads are worse. The American toad *Bufo terrestris americanus*, which produces a trill 1,500 Hz call, has an auditory threshold of 50 decibels, and green tree frogs and even poorer. To make up for this, however, male frogs and toads make extremely loud sounds. A male cricket frog one metre away can be heard to be pumping out a painful 115 decibels.

Some researchers have suggested that the rather dismal auditory threshold of many species may indicate that frog calls are mainly for short-range communication rather than signal designed to attract females to breeding ponds from a green distance. In fact, several studies suggest that females can locate suitable breeding ponds in the absence of calling males, perhaps by smell. Another aspect of Bob Capranica's work has involved the way sounds are processed and recognised in the nervous system. What is it that allows a female to recognise a call from a male of her own species even when there is cacophony of background noise masking the signal? In order to attempt an understanding of the auditory process, Capranica and his co-workers have developed apparatus which enables them to record from single nerve fibres in the auditory nervous systems of frogs and toads. An electrode connected to a single neuron of an anaesthetised animal can record its sensitivity to frequency, temporal patterns, sound thresholds, and sound pressures.

By recording from nerve fibres, both from the ear and in the central nervous system, an understanding may be gained of how complex sounds are being processed and recognised in the brain. One way in which sounds might be recognised is to specialise the ear itself, right at the beginning of the reception system. It would only be stimulated to respond to a call carrying the correct information. This might be called peripheral specialisation. Another way is to have an ear that picks up everything and then leaves it to the brain to sort it all out.

Humans, for example, have an auditory system sensitive to a broad band of frequencies. Our ear does not show any specific selectivity for particular kinds of sounds–we are sensitive to music, speech and the calls of many other animals. Selection takes place in the brain. Specialisation at the 'ear-end' means the brain has a simpler job, but at the cost of limitations in the kinds of sounds that can be heard. Bob Capranica has found that the frog's auditory system is specialised at the ear. The frog ear is tuned to detect, say, the mating calls of its own species, and not the mating calls or any other calls of other species. The peripheral specialisation is species specific, so the ear of a bullfrog is tuned to different frequencies than that of a tree frog.

In addition, in the cricket frogs, the ear of the female is turned selectively to the local dialect. A female cricket frog in central Texas is completely deaf to the calls of males from east or west Texas. A further degree of selectivity in the ear has been found in another tree frog from Puerto Rico, called *Eleutherodactylus coqui.* It has the specific name 'coqui' because the male produces a two-note call that sounds like 'ko-kee'. It produces a constant frequency 'ko' note, followed by an upward frequency 'kee' note. Capranica and his colleagues were puzzled to know why a male produces a two-note mating call, as most species studied in temperature regions of Northern Europe and America have only single note calls. Collaborating on this investigation was Peter Narins, now at the University of California at Los Angeles. It took them to the rain forests to Puerto Rico. There, tree frogs begin to call around dusk, and continue to call throughout the night until the early morning hours. The call, however, is not constant; it changes during the evening.

In the early part of the evening, the males set up their calling territories, each male returning to the same spot each night. They are aggressive and will sometimes come to blows if two individuals find themselves too close together. During this period of territory occupation the males produce only the 'ko' note. Neighbouring males will answer each other with this single note. After about an hour of vocal exchanges the males would add 'kee' to their calls, so the forest would echo to the sound of 'ko-kee'. In order to interpret the two signals Narins and Capranica once more employed playback experiments. A loudspeaker was placed near to a 'ko-kee' calling male and both natural and synthetic calls were played to him. If the entire 'ko-kee' call was played at relatively low intensity the caller paid no attention. He just continued giving his 'ko-kee' call one once every three seconds. If the volume of the playback was increased the caller would immediately turn to face the speaker, stop singing the 'kee note, and respond to the playback with the single 'ko' note. At the same time the caller would approach the loudspeaker.

By increasing the volume of the playback the researchers had simulated an approaching male; the louder the sound, the closer the supposed intruder. Narins and Capranica concluded that there must be some territorial communication going on between neighbouring males. If the 'ko' note alone was played at a sufficient volume to a 'ko-kee' caller he would respond by dropping 'kee' and calling back with 'ko'. If the 'kee' note alone was played, however, the frog would ignore it, no matter what the volume. The researchers then turned their attention to the female frogs. First, they played back 'ko-kee' calls to which the females immediately responded. They approached the loudspeaker. If the 'kee' note alone was played they would also be attracted towards the sound. If, however, the 'ko' note was played the females paid no attention and often hopped off in a different direction.

Clearly 'ko' is a territorial rivals. If two males get too close together, they are no longer interested in attracting a mate; instead they become preoccupied in solving the aggressive encounter with the neighbouring male. Most territorial disputes are resolved by calling. The males gradually separate and when the intensity of the call is sufficiently low, each male will resume its 'ko-kee' call is sufficiently low, each male will resume its 'ko-kee' call and attempt to attract a female. Here, then, is a simple communication system – the male produces two notes; one note signals to males, the other to females. It is possibly the simplest communication system for interaction between the two sexes. And, when the physiology of the auditory system is examined, it is revealed that the ear of the female is selectively turned to the "kee" note and the male's ear tuned to the 'ko' note. So, in the auditory system of frogs, Bob Capranica has demonstrated not only species specificity in the sensitivity of the ear, but also geographic and sexual specificity. Work by Peter Narins and Randy Zelick with *E. coquoi* has recently complicated the picture in that female tree frogs have been found to respond to sounds outside the frequency range of the species-specific call.

The biological significance of this is not clear, although it could aid the frog in predator recognition – the Puerto Rican screech owl *Otus nudipes* has been found to take *E. coqui* as 30% of its diet. It could also be that the broad auditory sensitivity allows the nocturnal frog to locate its own flying insect prey. *E coqui* is also quite capable of detecting the first advertisement notes of a related species, *E. partaricensis* which lives in the same vicinity. The researchers propose that our interpretation of species–specific calls of frogs and toads should be modified. In order to examine the next stage in the hearing process, namely how the information is processed in the brain, Capranica turned his attention to the larger bullfrog. An analysis of its mating call revealed there were three peaks of energy, analogous to human vowel sounds; a low frequency peak arounds 200 Hz, a trough of low intensity at 500-700 Hz, and a second peak at 1,400 Hz.

It was found that, in order for a bullfrog to respond to the mating call, the two peaks of energy must be present. If the low frequency peak is filtered there is no response; likewise with the high energy component. 'How then,' asked Capranica, 'are these two energy bands actually recognised by the animal; what is the physiological basis? The frog ear differs in one respect from that of birds, mammals or reptiles. In humans, for example, a singled auditory organ, the cochlea, converts sound signals into the nerve impulses that go to the brain. In frogs and toads, though, there are two organs in the inner ear, the amphibian papilla and the basilar papilla. The amphibian papilla is tuned to frequencies between 100 Hz and 1,000 Hz, while the basilar papilla is tuned to 1,400-1,500 Hz. For the bullfrog to recognise its mating call, the amphibia papilla must be excited by the low frequency peak in the call at about the same time as the basilar papilla is stimulated by the high frequency component. If other is missing the sound is not heard. The low intensity mid-frequency component mechanically suppresses any excitation of the amphibian papilla at 500 Hz so this papilla only responds to the 200 Hz peak. Other calls excite different combinations of the two papillas. The distress call, containing only high frequencies, triggers the basilar papilla only. The release call is in the mid-frequency and high-frequency ranges, and so stimulates nerve fibres from both auditory organs. What then happens to the information when it reaches the brain?

Bob Capranica has postulated that the brain of the frog has particular detector sites for each call–a mating call detector, a distress call detector, a release call detector, etc. dedicated sets of nerve cells in different parts of the brain detect particular sound patterns. Stimulation of those cells by the appropriate combination of frequencies and temporal spacing would somehow lead to an appropriate behavioural response. Karen Mudry, then working with Capranica, but now at the University of Akron, Ohio, identified two auditory centres in the frog brain–one in the brain stem, in the thalamus, and the other in the telecephalon, which corresponds to the human cerebral cortex. In these two centres. Mudray discovered the organisation that had been postulated would be required for a mating call detector. She found an auditory centre that would neither respond to low frequency sounds alone nor to high frequency sounds alone, but which would respond when the two were presented together.

Furthermore, the centre would only respond when the temporal features of the sound matched the energy peaks in the mating call. Each bullfrog croak lasts about one second, and is separated from the next by an interval of about seven-tenths of a second. The number of croaks given is variable depending on the maturity of the frog. Typically, a male will give seven or eight croaks and then fall silent for about a minute. It will then repeat the series of croaks. Each croak has a gradual onset time in that it takes about one-tenth of a second for the croak to reach maximum intensity, eight-tenths of a second of constant intensity, and then a gradual decay lasting one-tenth of a second. In the auditory centre found by Mudray, the acoustic envelope required to stimulate that centre must have a rise time of 100 milliseconds, as in the natural sounds. The work continues to track down a similar detector in the frog brain for distress and release calls. But, could this amphibian model have wider implications? Is this the way a brain recognises a complex sound? Or is the simple notion of a detector in the brain for sound naïve?

Female Choice

'Explosive breeders', such as the European common toad Bufo bufo and the common frog *Rana temporaria* do not seem to 'choose' mates. Males and females arrive simulataneously, in large numbers, at ponds whereupon the males grasp any moving object and hope it is female. The common toad does not have a mating call as such, relaying on the relying on the release call for ex recognition. Females have very little opportunity to select individual males because they are often grabbed by a searching male during migration to the pond. Male European common frogs do have a clearly audible call but playback experiments have not shown that this is excite the female and encourage her to spawn. Many paired males have been heard to call as well as single ones. In the African clawed toad *Xenopus laevis*, the playing of recorded male calls alone to single females is enough to cause spawning. In many other species of frogs and toads females do the choosing and recently researchers have become interested in the way a female discriminates between individual males of her own species.

Some males, it seems make better mates than others. Females might prefer to mate with older males because age itself is a good

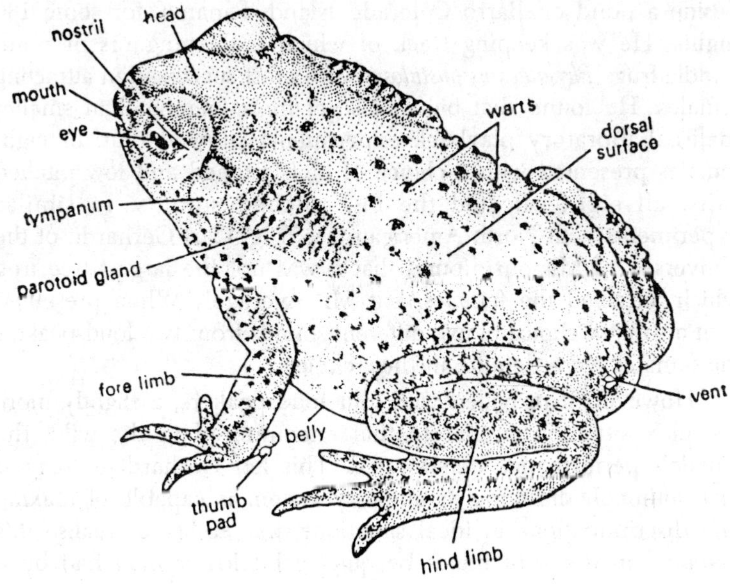

Fig. 2.3. Bufomelanostictus.

indicator of a male's ability to survive over many years, and therefore an indicator of good quality genes. It is becoming apparent that females are able to select mature males simply by listening to their voices. When a female enters a pond of calling males during the mating season she swims around, apparently randomly, seemingly listening to the cells of the males. After a while she will select one individual, head towards him, and eventually spawning will place. What features are in that make's voice which are preferred over those of the calls of all the other males? Why does the female head for the caller, rather than a neighbour? Could it be the pitch of the call which is important? In many species of frogs and toads, as the male matures and becomes older becomes larger. As he gets larger, so too does his vocal apparatus. Increase in the size of the vocal cavities gives rise to louder calls, while increase in the size of the vocal chords results in deeper pitched croaks due to a lower frequency of vibration of these structures. In humans, the same happens.

As we go through adolescence the pitch of the male's voice drops. It appears that in some species of frogs and toads females are more attracted by a call of lower pitch than one of higher pitch. Michael Ryan, now at Cornell University, New York, sat beside a pond on Barro Colorado Island, Panama, for some 180 nights. He was keeping track of which male tungaras or mud-puddle frogs *Physalaemus pustulosus* were most successful in attracting females. He found that big males get more mating than smaller males. Laboratory playback experiments showed that, of eight females presented with a choice of high pitched and low pitched calls, all eight selected the low pitched sounds. In similar experiments with North American tree frog, Carl Gerhardt, of the University of Missouri, found that it was not the largest tree frog which attracted the females, but Mr. Average". When presented with a straight choice of two different sounds from two loudspeakers the females were definite in their choice.

However, if provided with four loudspeakers, a slightly more complex situation similar to that encountered in the wild, the female's performance deteriorated. This led Gerhardt to suggest that, although the female's auditory system is capable of making fine discriminations in ideal situations, her ability to realise this potential in the wild might be quite a bit lower than had been found in the laboratory. Gerhardt went on to postulate that female green tree frogs prefer 'average-frequency-calls' because of the presence of two other species of tree frogs which have mating calls above and below the frequency range of green tree frogs. The barking tree frog *Hyla gratiosa* calls with a slightly higher pitched call. If female green tree frogs were to prefer a particularly high or low call from the male it might accept a mate of the wrong species, producing hybrids of low viability. There are constraints, it seems, on how' choosy' females can afford to be in nature. Anothony Arak, of Cambridge University, carried out similar frequency discrimination test with European natterjack toads *Bufo calamita*, an amphibian on the British is of endangered species, and one of the few in the British Isles that makes any sound at all. Female natterjacks were exposed to playbacks of high and low pitched sounds and showed no preference at all. Arak then gave them a choice of quiet and loud calls, and generally the females chose the louder calls. This was an interesting result for loudness, too, can be correlated with maturity and body size. Arak also played back the calls at different repetition rates and found that females preferred calls at faster rates, although under normal circumstances they are unlikely to encounter such variations.

In the wild, louder or faster calls might not actually attract the females. Laboratory experiment results may simply represent a super-stimulus effect where, if a stimulus is produced at a higher level it is preferred to one at a lower level. Anthony Arak went on to see if the laboratory observations of female choice coincided with those made in the wild. He looked at which males in the population obtained the most females. Large male natterjack toads in a population were seen to obtain as many as eight females during the breeding season. Much smaller males obtained fewer females. Arak tried to find out why. He measured different aspects of the male's calls, whether it was present at the pond on some nights and not others and so on. He soon found that male natterjack who turned up at the pond on more nights were most successful, although these males were not larger than average.

However, he also found that, on any given night, larger males were more likely to mate because they called more loudly and more frequently than the smaller ones. The larger males, though do not have everything their own way. Some smaller individuals do gain access to the females, and they have gone about it in a rather clever way; it is called the satellite or 'sneaky' male strategy.

What happens is that the smaller males wander around the pond until they encounter a large calling male that is likely to attract the females. The smaller frog sit in its immediate vicinity and remains silent. If the little frog called it would be chased away by the bit one. When the female approaches the resident large male, the sneaky satellite male nips out and intercepts her, grabbing, her in amplexus before she reaches the larger male. In ponds, often only one-third of the male frogs call, the rest remain silent. Sometimes the large frog spots the satellite male as it reaches the female and a fight ensues.

A satellite male, though, once firmly grasping the female is different to dislodge. Females seem to avoid satellite makes, if at all possible. Female tree frogs sometimes shake the little sneaky below the satellite male. Surfacing nearer to the territorial caller. Any toad not calling that approaches a female natterjack toad will cause it to swim rapidly towards the centre of the pond. Females may avoid satellite males for a variety of reasons. May be they are genetically interior. They is also the danger of mating with the wrong species. Since a satellite male it silent the female does not know if he is a male of her own species. Sometimes several species breed alongside each other, for example, common toads and natterjack toads. Female natterjack must be ware of the promiscuous 'explosive- breeding' male common toads, which grab anything that moves. Males do not have to be so fussy.

A male toad has little to lose through a hybrid because he can always try again. The female would have spent a year feeding in order to mature her eggs. A hybrid mating would result in non-viable spawn and her efforts would be wasted. Silent males are not silent all the time. If a calling male is removed the neighboring satellite male may choose to start up. It is also possible to change a calling male into a silent satellite male by plying his call back at a louder level; when outshouted it seems best to shut up. In the case of tree frogs, calling perches are at a premium. The best places are aggressively defended. If kicked off or unable to obtain a perch, a satellite relationship with the frog in residence is the next best strategy to adopt.

Satellite tree frogs studied by Carl Gerhardt in North American were equally successful to calling males at obtaining females. Satellite males, in the main, tend to be young and small creature incapable of producing loud calls, and therefore obviously less attractive. If a female is going to discriminate against such a caller it is best for him not to utter a word until he reaches maturity. The satellite strategy, although less productive in some species (satellite natterjacks do badly, while satellite green tree frogs do as well as callers) is less expensive in energy terms. Calling is restructured almost entirely to the resources. Body fat reserves have been shown to be depleted considerably after a heavy fighting and calling period. Calling, though, is energetically less expensive than fighting, but is still about six times as expensive as sitting around or feeding. In North Carolina one enterprising species of toad appears to have developed another form of 'sneaky' strategy to acquire a mate. Small toads have been reported to have been seen sitting in the cooler parts of ponds in order to lower the pitch of their voices. By positioning themselves there, the young pretenders mimic their deep calling rivals.

Deceptions such as this amongst the Fowler's toads *Rufo moodhousei fowleri* of North America are the first known cases of amphibians using surrounding temperature to make themselves more appealing to potential mates; at least, according to Lincoln Fairchild, of Ohio State University. Other herpetologists have questioned the work. Temperature can change certain properties of a call, the work. Temperature can change certain properties of a call, including pitch, but in a very insignificant way. It would need to be a large drop in temperature variations on the calls of some species and the largest change so far found is in the order of one to tow percent change in frequency per degree Celsius change–a small amount. Frogs, though, are poikilothermic or cold blooded animals. Their body temperature varies with he surrounding air temperature.

In temperature latitudes, therefore, frogs call in the early evening, when the pond is still warm, yet visibility is deteriorating for predators. Most species call for the first few hours after dusk and then, as the night draws in and the air gets colder, they stop calling altogether. I the tropics, where temperature may not vary, frogs and toads will call throughout the night. A drop in temperature will slow down the metabolic rate of a frog. It simply becomes more lethargic, which mean it is unable to call a quickly or a effectively. The pitch and temporal patter will vary. Female frogs and toads, have overcome this problem may locking onto the males temperature frequency response. As the temperature fluctuates, so the preference goes in favour of higher or lower pitched call, whichever is appropriate. No matter what the temperate the females always knows at what frequency a male of her particular species should be calling. In the Canary Islands, the male Mediterranean tree frog *Hyla meridionalis* gives a call which is temperature dependent and the female respond preferentially to certain temperature.

Hans Schneider, at Bonn University, has been analyzing there calls and has found that the optimum temperature at which the frogs do their calling is between 12° and 24°C. At 24°C a call is brief, about 300 milliseconds, and the interval between calls is just one second. Each call consists of 37 impulses of sound produced at a rate of 125 impulses per second. As the temperature drops, as night falls for instance, the rate changes. At 12°C the call lasts about half a second and the interval between calls extends to 3 seconds. Each 120 call has 43 impulses which are delivered at the slower rate of 90 impulses per seconds. In playback experiment with artificial call, in which the parameters of the calls could be altered. Schneider found that females did not prefer callers with the same body temperature, but would move towards caller with a slightly higher temperature. As the outside evening temperature fall, the males sitting in the waste lose less hear (and indeed, produce excess heat during calling), than the females approaching the pond overland. A female suggest Schneider, move preferentially to a male which, through its call is indicating that it is at a slightly higher temperature than the female and is therefore probably in the right environment i.e., the pond, for mating to take place successfully.

Male Competition

The common toad *Bufo bufo* in Britain is unlike many other frogs and toads in that does not often use a call to attract females. There is some evidence that in this and a few other species the mating call has been lost in the course of evolution. It has been suggested that because common toads always go back to the same ponds, year after year, even the extent that there are records to toads returning to a place where the pond has long since vanished, they don't need calls to tell them where they should aggregate to breed. If they are all heading for the same place, they will meet up anyway. In early spring huge numbers of toads can be found migrating to their breeding ponds. The invariably take the same route, and at major roads ' toad-crossing' patrol now ensure their safe journey. Whether the common toad has actually lost its call during evolutionary history, or whether it never had one, is not clear.

Some toads do make a soft croak in addition to the release call, but there have been no studies to establish whether this is a mating call. There have been studies on another call, a soft peeping sound, which males use in the presence of other males and which functions as a special form of release call. The studies were made by Tim Halliday, from the Open University, Milton Keynes, who was interested in working out which toads, males outnumber females are about three to one, and Halliday wanted to see whether large or small male toads had females and which size kept its prize. He found that the lager males tend to hang on to their mates while the smaller ones were often displace. This led Halliday to examine the way these toads fought for dominance. A female approaching the pond is intercepted by a male. He leaps onto her back, grasps her behind her front legs with his, and holds on.

Any other male interested in taking his place will need to force him from his position. A large male, after a prolonged fight, is likely to dislodge a smaller toad a fight might last for six hours or more. During the fight the males produce a soft peeping sound, but only when there is actual physical contact between them. Halliday wondered whether these calls were some how involved in the fight procedure; perhaps the call is saying something about each of the males, may be about their relative sizes. If a large male can indicate to a smaller one that is more powerful and inevitably will win a fight, it may induce the small male to move aside and stop fighting. Halliday and his colleagues recorded the calls from a variety of males of different sizes and noticed that there was a direct relationship between the pitch of the call the size of the toad. Large toads had low pitched calls and vice versa. Did they use this information in fights? To test for this the researches muted a medium-sized male toad on a female's back by passing a rubber band across its mouth and under its armpits, like a horses bit, so that it couldn't produce calls of its own. Recorded calls were then played to the pair in amplexus.

The researchers then waited for a second male to attempt to dislodge the muted male. When the second male attacked the recorded sound were played. The attack was much intense if the high pitched call of a small male were played than for deeper pitched calls. Calls. The attacker appeared to be responding, not to the actual size of the muted male, but to its apparent size on the basis of the call. That, though, wasn't the complete story, for the physical strength of the defender was apparently being registered by the attacker. The attacker was receiving a collection of cues, of which the call was but one. Many tree frogs give encounter calls which seem to act as a kind of threat. If a defending male on a prime singing perch on a bush or shrub is approached by another male, it will cease producing a mating call and begin an encounter call, a pulsed trill. If the aggressor does not pull out a wrestling match ensues. Eventually one frog will fall off the perch and the other, the winner of the bout, begins to call again. Anothony Arak has been studying the tree frogs in Sri Lanka.

In one species *Philautus leucorhinus,* the male defends a perch on prominent blades of tall grass. When two males get close they give a pulsed threat call and then proceed to match each other's calls in an escalating singing contest which may last for ten minutes or more. If neither male backs down, the event ends in a fight. Arak could find no pattern related to male body size that correlated with the way residents were or were not supplanted. He compares two call matching male tree frogs to two bidders at an auction. They progressively increase the number of note in their calls in response to a rival males's last bid. It is not clear, as yet, whether they are bidding for females sitting near the calling perches or simply bidding to secures their perches. Male tungara frogs do not have a territory in the classic sense, but they too, interact with each other. They use to call which sounds a little like 'ee-y-ur', and get so upset that they turn and face away from each other.

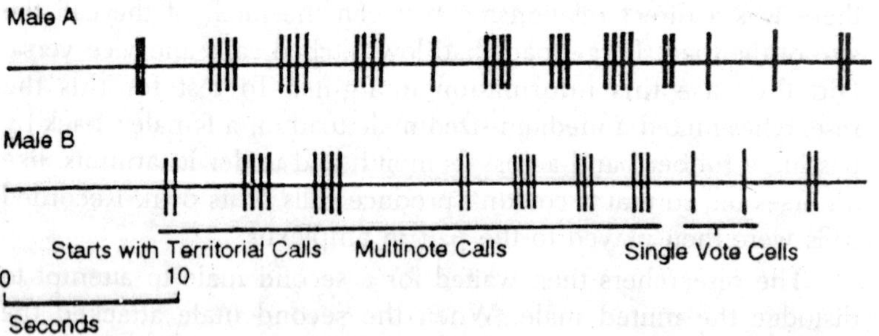

Fig. ? 4 Male tree frog vocal interactions.

These miniature frogs, the size and shape of a square ping-pong ball with tiny legs sticking out a the four corners, back towards each other, kicking fiercely with their hind legs sticking out at the four corners, back towards each other, kicking fiercely with their hind legs, until one gives up and goes away. Another group of tree frogs from Panama have complicated two part calls similar to those of the tungara frog; introductory whistles are followed by a series of click like notes. *Hyla ebraccata,* for example, has a long introductory note, 'eeeeer' followed by may be two sometimes four, clicks: eeeeer-i-i'. Males calling by themselves give only the introductory note.

Only in competition with other males is the staccato series of clicks produced. Essentially it is a competitives signal. What one male emits determine what another produces. Kentwood Wells, of the University of Connecticut, has been studying these and other *Hyla* species of tree frog. Originally he thought that these frogs may be matching calls, much like Arak's Sri Lankan tree frogs. If a male gives two clicks then the neighbour may replay with two. There could be an escalation of clicks, and female arriving on the scene might compare the would-be suitors and ;pick the frog with the highest number of clicks. Wells believe now that this is not what is going on.

Further study has revealed that frogs overlap their calls with those of their neighbours. When a male hears his neighbour begin to call, he waits for a set period and then starts to sing his own song. The neighbour's introductory notes are clearly heard but his terminal clicks are masked by the second male's introductory notes. This is an interference strategy. If the clicks are attractive to females the second male makes it difficult for the female to home in on his rival by masking the attractive part of his call. In addition, the second male may give enough click notes for his call to continue well after the rival has finished. May be the last male frog to be heard gets the female. Many researchers have speculated that, in frog choruses, there is a dominant male who starts things off, and that this male is the most attractive frog in the group.

Whitney and Krebs studied calling in the Pacific tree frog *Hyla regilla.* These little frogs congregate at breeding ponds and sing in choruses. There are calling periods alternating with periods of silence. Each bout of calling, though, is initiated by a bout leader, which is not only the first to sing, but also the last to stop.

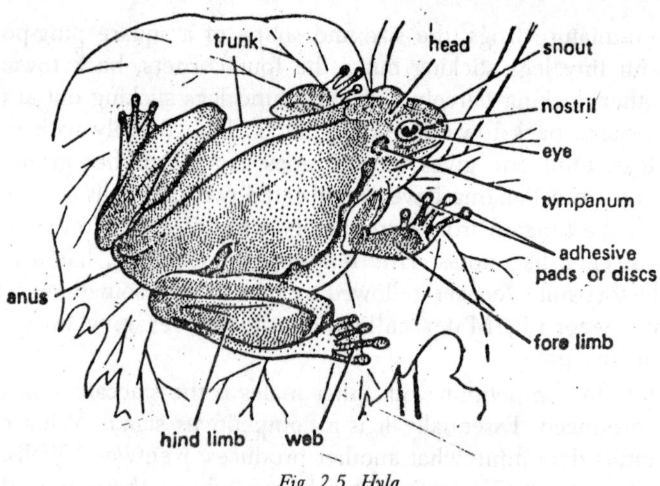

Fig. 2.5. Hyla.

About leader may call faster than the rest of the chorus and is more likely to break out into a solo between choruses. In one experiment, a female was placed in an arena between four loudspeakers, each playing back identical calls. One loudspeaker was made to start and stop the chorus of four. In 14 out of 18 playbacks the female headed towards the 'leader' loudspeaker. Whitney and Krebs had shown that the frog which starts and finished is the most attractive to females. Kentwood Wells' argument goes even further.

Simply on acoustic grounds the last male to call may be the most perceptible to the female and therefore attracts the female for successful mating. Wells emphasises that this is only an hypothesis, yet to be tested. In a tropical forest the sounds created by frogs can be deafening. A person would have to shout loudly to be heard above the din. As often a not, there are many species of frog calling together. The problems each species has in hearing its own calls are immediately obvious. For example, calling in frog choruses with *H. ebraccata* is usually *H. microcephala* and *H. phlebodes.* Unfortunately all these tree frogs broadcast in the same frequency band. It is as if several radio stations were all playing on the same frequency and interfering with each other. All three frogs can hear components of each other's calls. If an individual male calls for a female how is she to pick out its call from all others?

Well discovered that *H. microcephala* call lasts for about 20 seconds, so *H. ebraccata* males wait for the periods of silence between

20 second calls and sing their songs in the gap. Wells has demonstrated that all these species of tree frogs are not sitting in the forest in isolation but they hear each other, respond to each other's calls and this determines how they will communicate each others of the same species. Most work until now has been concerned with the way calls which differ in temporal pattern or frequency structure serve as an isolatiòn mechanism to ensure that a female always responds to the call of her own species. They ways in which a species overlaps or interferes with another have yet to be studied. Given that many frog choruses in the tropics have may be ten or 15 species of frogs calling, interspecific interference is likely to be widespread. Wells feels that research might usefully turn its attention to similarities in calls, for instance, rather than differences.

The Story of the Frog and the Bat

A male frog's call may not only attract females or repel rival males, but will also draw the attention of predators. Snapping turtles, for instance, are known to home-in on the calls of buildings. For the mud-puddle frog on Barro Colorado Island, danger swoops in from the air. The aerial predator is not a hawk or heron, but a frog-eating bat. The mud-puddle frog of the rain forests is a miniature animal about in inch long. It has attractive mating call which, unlike those of many other frogs and toads, is not just a monotonous call; it has a variety from which to choose. A single male, by itself, goes 'aow, aow' and so on, making up to 7,000 calls in one evening between sunset and two a.m. It is using the call to attract a mate. If, however, there are several males within earshot–there may be ten males in a pond three feet long by a foot across– then the caller produces the sound 'aow-chuck' or 'aow-chuck-chuck' and so on up to as many as five or six 'chucks' at the end of the line. Stanley Rand of the Smithsonian Institute, Washington DC, was intrigued to know that functions of the elements of the two types of call. He made playback tapes–one with the 'aow' call and another with 'aow-chuck-chuck'. Females were then released into an arena and the two calls were played. If 'aow' was played, the female would move towards the loudspeaker. She would also approach if 'aow-chuck-chuck' was played. Rand edited out the 'aow' segment leaving the 'chuck-chuck'.

The female was not interested. If, however, she would choose the latter. But why, thought Rand, did a male which had gone to the trouble of calling 7,000 times in an evening ever give anything that was less than maximally attractive? The 'chuck' part of the call, it turns out, contains a broad band of frequencies, is hard-edged, and therefore easily locatable. By adding the 'chucks' the male would simply make himself easier to find; or so it seemed. Also working at Barro Colorado was Michael Ryan. It was he who noticed that females prefer male frogs with deeper voice. In further tests he showed that females could, in reality, locate a male's call without 'chuck' just as well as a call with 'chucks'. He reasoned then that the 'chucks' contain the preferred drop in pitch which goes with larger males. Ryan carried out more tests. He prepared a series of recordings where the pitch of the 'sow; remained constant and the pitch of the 'chuck' was varied. Given a choose of high or low pitched' chucks, the female chose the deeper sound and thus the larger frog. So the females are, indeed, attracted to a male providing size information in its call, i.e, one giving deep pitched 'chucks'.

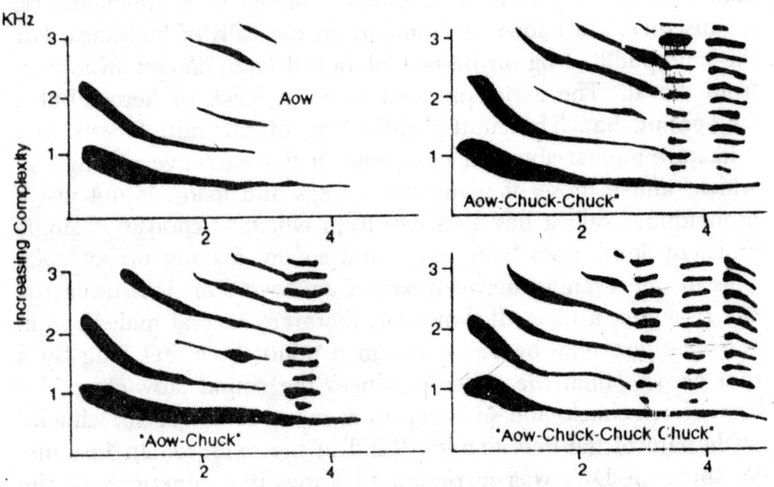

Fig. 2.6. Calls increasing complexity of the mud-puddle frog.

The disadvantage though, of adding this locatable cue is that it make it easier for a predator to locate the called too, and it was not until bat researcher Merlin Tuttle, of the Milwaukee Public Museum began studies on Barro Colorado Island that a predator was identified. One day he caught a bat with a frog in its mouth. It was the fringe lipped bat *Trachops cirrhosus*. Merlin Tuttle and Michael Ryan went on to unravel that story of the frog and the bat. The first question Tuttle and Ryan addressed was, might predation play some role in the evolution of the complex vocal repertoire of mud puddle frogs? In particulars, do male leave out 'chucks' at the ends of their calls some of the time because of the danger of attracting bats? Their first experiment showed that the bats do respond to the sounds of the frogs. Tuttle and Ryan went out into the forest and using a night vision scope, watched the frogs at their breeding ponds. They could see that the bats fed quite frequently on the calling frogs, swooping down and pulling the frogs out of the water. They also noticed that the hunting success of the bats was much higher when the frogs were calling than when they were silent. Back in the laboratory, Ryan carried out a series of play back experiments using a league flightcage. A bat was captured and then released into the cage. Frog calls were played through a loudspeaker.

By using the speaker, the researchers had eliminated the shape of the frog as a visual cue for the bat and had also ruled out the use of echolocation. When the frog calls were played the bat flew directly towards the speaker, of ten ripping at the experiment in the forest. Through the night vision scope they could clearly see bats heading for the loudspeaker attracted by the calls of the frogs. Bats, however, have an auditory system adapted to ultrasonic frequencies; the middle ear apparatus for example has bones reduced in mass so that they vibrate more easily at very high frequencies. How can a bat detect the low frequencies, to fringe lipped bats and found that they were sensitive to the frequencies at which the frogs were calling. Although this species of bat has an auditory system sensitive to is own echolocation call, it is also capable of hearing lower frequency sounds. The same may well apply to other species of bats. But there are many kinds of frogs and toads living in and around the ponds of Panama; some potential food for the bats, other poisonous, could the bars pick out one frog from another simply by their calls?

In further playback experiments, bats were given a choice of calls. One speaker played the calls of a species of frog known to be predated by bats, while the other played the calls of toads of the genus *Bufo*, all of which have poisonous skin secretions. The bats preferred to attack the speaker playing the call of the edible frog, and for the most part ignored the speaker with the call of the poisonous toad. A similar experiment was carried out with the

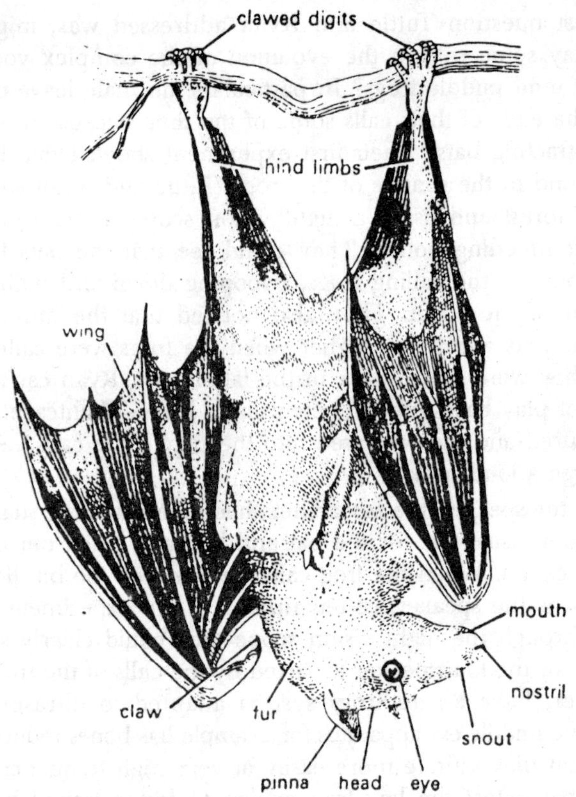

Fig. 2.7. Pteropus medius (Flying fox).

calls of the South American bullfrog. *Leptodactylus pentadactylus.* Some individuals of this species grow to a foot in length from snout to vent and are much too large for the bats to capture. Occasionally these giant bullfrogs will turn the tables on bats by leaping out of the were and capturing them as they fly by. Ryan gave bats the choice of the calls of bullfrog and their preferred edible frogs. They avoided the bullfrogs. Evidently bats are able to distinguish one species of frog from another purely on the basis of the call. If the frogs, then are to take evasive action they must be able to find out when an attack is coming. It bats can detect frogs, can frogs detect bats? It seems that they can. Turtle and Ryan recorded a large chorus of frogs and noted what happened to the calling behaviour before and after a bat flew over. If it was a moonlit night or a night with no moon but no cloud cover, the frogs would stop singing immediately a bat flew across the pond.

On a totally dark moonless, cloud night, the frogs showed no response to a passing bat and continued calling. By controlling light intensity, stimulating light and very dark nights. The researches also eliminated the bats echolocation calls as a warming cue for the frogs. High frequency sounds, similar to those emitted by the bar, produced no response from the frogs. Two species of frogs on Barro Colorado Island form the main part of the diet of fringe lipped bats; one is the mud puddle frog and the other pugnosed tree from *Smilisea sila*. Tuttle and Ryan have found that there are tow quite different, almost opposite, strategies, to avoid capture, by bats. The mud puddle frogs gather in large groups and sing in chorus. If there is a constant level of perdition, the chances of being eaten are relatively low if an individual is surrounded by many others–the 'selfish-herd' effect.

The pugnosed tree frog, on the other hand, does not call in choruses by is spaced out every five meters or so along stream banks. It too, can spot the approaching bar, but behaves quite differently according to whether the night is bright or dark. On bright, moonlit night the frogs sit and call out in the open, on the tops of rocks and remain there for much of the night. On nights without a moon, when they are unable to see the bats flying down the stream they start to call at dusk but then quickly move to the tops of trees, or stop calling altogether and hide under rocks, fallen tree trunks or large leaves. They decreases their exposure to the bats when they cannot detect them.

3

COOPERATIVE BEHAVIOUR

The African elephant (*Loxodonta africana*) may have one of the most advanced of all mammalian social systems (Wilson, 1975). For elephants, group living provides individuals with defense from predators and enhances their ability to locate resources. Although group membership offers the advantages of cooperation, it also entails the disadvantages of competition for scarce resources and breeding opportunities. Research on African elephant social organization has emphasized female cooperation but neglected the obvious potential for intersexual competition. This chapter reviews a number of reported field observations of cooperative and competitive interactions among herd females and identifies actors that may influence the development of elephant behavioural patterns. Along with unusually long lifespans and broadly overlapping generations, elephants have developed a complex, matriarchal society (Douglas Hamilton and Douglas-Hamilton, 1975; Moss, 1976).

Elephant herds comprise groups of related females and juveniles and calves of both sexes. The herd's activities are directed by the largest elder female, the matriarch. Relative social status of individual within the herd is determined by kinship lines, age, size, and, Individual disposition; a rank order hierarchy results. Female offspring remain in their maternal family units or larger kinship groups throughout their lives. This promotes long-lasting bonds between females. Young males, on the other hand, are forcibly excluded from maternal herds at puberty. Their subsequent associations with female relatives and natal herds may be common but are currently not well understood. Adult bulls spend most of their lives alone or in short- lived, loosely organized bachelor herds, Joining female herds only intermittently, usually for breeding. Elephant societies are complex and embody a dynamic balance of affiliative and agonistic interactions between individual herd members.

All individuals share the common goals of survival and reproduction, and their cooperative and competitive behaviours are directed towards this end. The behavioural patterns displayed by a specific individual may be determined by a set of factors including age reproductive state, and physical condition, all of which contribute to the animal's status in the breeding hierarchy. These variables change over time, altering the relative benefits of cooperation and competition for the individual (Chase, 1980). The behaviours expressed may also be directly influenced by environmental factors, including the quality, dispersion, and abundance of forage; water and mineral availability; and predatory pressures. The balance of this chapter reviews two basic elements of herd social structure: cooperative and competitive Interactions between females holding unsocial status. Section II reviews the long-recognized cooperative behaviour of herd females. Group defense (Section II, A) and resource acquisition leadership (Section

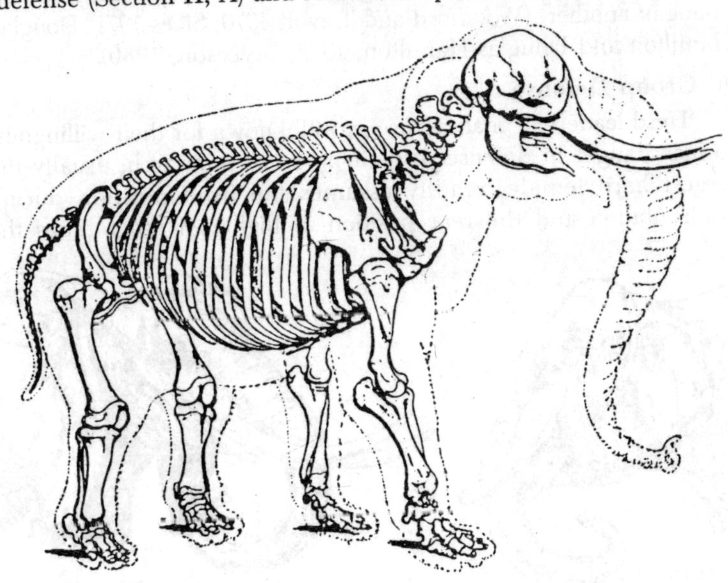

Fig. 3.1. Skeleton of Indian Elephant.

II, B) are discussed first. Then, allomothering is emphasized, and several types of benefits-and costs-to individuals that might result from this behaviour are explored (Section II, C). Section III discusses competitive interactions in elephant herds that would affect the reproductive success subdominant females.

A discussion of the importance of seasonal timing to reproductive success (Section III, A) precedes an exploration of the relationship between social status and breeding opportunity (Section III, B). Social interactions that may affect differential reproductive success among females of unequal rank are explored (Section III, C). This section does not attempt to encompass all the types of competitive interactions that may exist in the elephant herd hierarchy. The dearth of research on this topic requires that even this limited discussion be speculative and conceptual.

II. Cooperation

Cooperative behaviours among cow elephants involve group defense (Sikes, 1971; Douglas-Hamilton and Douglas-Hamilton, 1975), locating of and leading to food, water, and other resources (Crook, 1971; Leuthold and Sale, 1973; Douglas-Hamilton, 1973, Hanks, 1979), and allomothering (i.e., one female's caring for the young of another) (Woodford and Trevor, 1970; Sikes 1971; Douglas-Hamilton and Douglas-Hamilton, 1975; Bryceson, 1980).

A. Group Defense

Herd leaders or matriarchs are well known for their willingness to face danger in defense of their herds. A matriarch, usually the largest herd female, readily assumes the front position during confrontation and the rear position in flight. At other times the

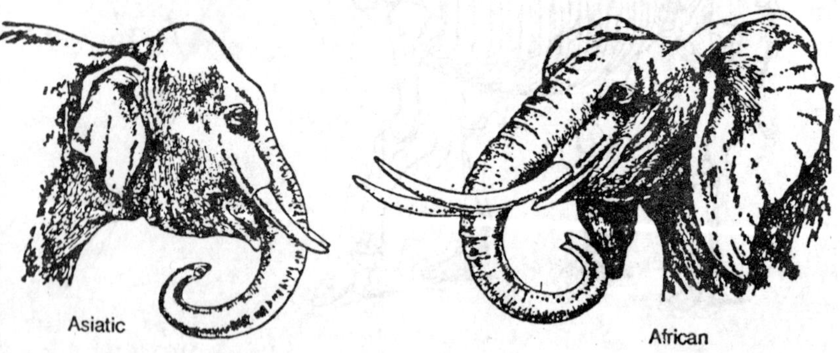

Fig. 3.2. Comparison of the heads of Asiatic and African elephants.

leaders arrange their herds into a circular defense posture (Kingdom, 1979). This defensive, formation protects younger members of the, herd by forcing them into the centre, a strategy found in other large herbivores, such as the musk oxen (Tener, 1954) and cape buffalo (Sinclair, 1977). The apparent altruism of herd matriarchs may be explained in several ways. Since the young they guard are likely to be close relatives (Douglas-Hamilton and Douglas-Hamilton, 1975), group defense should enhance their inclusive fitness (Hamilton, 1963,.1964) through kin selection (Maynard Smith, 1964). Matriarchs may also gain from reciprocity (Trivers, 1971; Axelrod and Hamilton, 1981) if, as a result, the offspring of females who guard others receive help in the future.

B. Fcmale, Leadership

In societies of long-lived and closely bonded individuals, leaders am often repositories of critical information about the surrounding habitat and its changes over both the short and long term. Douglas-Hamilton and Douglas-Hamilton (1975) and Leuthold (1977) suggest that the matriarchs many years of experience help them guide daily and seasonal movement patterns in search of food, water and mineral resources. Most areas in which African elephants evolved are characterized by highly localized, seasonal rainfall. Prime wet and dry season feeding and watering grounds may be widely dispersed within the home range of a family unit. Consequently, elephant herds show distinct wet and dry season ranges that may vary tremendously in size depending on local habitat conditions, and human habitation patterns. Within these ranges, elephant families do not show random utilization patterns. Rather, they display a great preference for some areas over others.

Movement patterns may be consistent from year to year, with individual herds returning to specific wet and dry sawn sites repeatedly (Leuthold and Sale, 1973; Douglas-Hamilton and Douglas-Hamilton,1975; Dublin, unpublished data, 1974-1976), or may be flexible in response to proximate environmental changes (Gorman, 1981). Female cooperation and coordination allow efficient herd movement to these resources over sporadically used routes. Nutritious forage and daily water are critically important to cow elephants, who gestate or lactate over half of their adult lives.

In fact, any relevant, learned information about the surrounding environment will be beneficial to the individual and the family as a while, including knowledge of potential threads posed by man and other predators (Leuthold, 1977, Curio et al., 1978a, 1978b). The important of this Information makes older herd females central to the survival of future generations. Long generation times, and the substantial overlap of individual life spans creates ample opportunity for the transmission of learned movement from old to young. In-elephants, this may include learned responses to such long-term recurrent events as drought Leuthold and Sale, 1973; Leuthold, 1977) and fire (Vesey-Fitzgerald, 1972). A herd leader's ability to locate and exploit new or spatially and temporally dispersed resources clearly benefits all group members.

C. Allomothering

Allomothering is common among African elephants (Sikes, 1971; Douglas-Hamilton and Douglas-Hamilton, 1975, Moss, 1976). Though older parous females will act as allomothers, young nulliparous females, usually assume this role. Allomothers direct calves movements and assure that calves are not left behind when the family moves away from resting or feeding sites. Surprisingly, lactating mothers also allow suckling by calves other than their own (Douglas-Hamilton and Douglas-Hamilton, 1975; Leuthold, 1977; Kingdon, 1979; Bryceson, 1980; Dublin, unpublished data, 1974-1976). The behaviour of allomothers toward their surrogate calves is often so intensely "motherly" that their true relationship is frequently unclear to observers (Kingdon, 1979). By assuming many mundane care responsibilities, allomothers free true mothers to forage at greater distances from the herd and allow them to suckle their young less often. The temporal stability of families and the overlapping nature of long generations assure herd females ample opportunity to interact with one another, thereby providing a favourable climate for the evolution of allomothering m elephant societies. This apparent family altruism parallels the associations described among cooperative breeders in man long-lived avian (Emlen, 1978; but see Koenig et al., this volume) and mammalian species (Kuhme, 1965, Schaller, 1972; Bertram, 1975,1976; Rasa, 1977; Frame et al., 1979; MacDonald, 1979; Moehlman, 1979; Rood, 1978,1980). The demographic portrait drawn by Brown (1974) for cooperative breeders includes traits such as low fecundity, delayed sexual maturation, long life spans, low dispersal, and recruitment of mother's helpers through retention of young. These characteristic traits are evident in several bird species as well as in elephants.

However, cross-species generalizations about the determinants of cooperative breeding can be misleading. For example, in some avian species, fledged birds only remain m their parents' territory and help in the rearing of subsequent broods when environmental conditions make their successful dispersal unlikely. In elephants, however, female young usually remain in their maternal herds and help with the care of younger group members.

1. Potential Benefits of Allomothering

Widespread and perpetual allomothering among elephants suggests that benefit is realized by one or all of the individuals involved. But what do allomothers gain relative to others who might not assume such energetic burdens? By tending to young other than their own, allomothers stand to increase any or all of the following; (a) inclusive fitness, (b) likelihood of reciprocation in the future, (c) experience in rearing offspring and (d) access to resources.

(a) Inclusive Fitness

A potentially important force behind allomothering is the opportunity to enhance ones inclusive fitness through helping close relatives (Hamilton, 1964). Assessment of fitness gains depends on accurate determination of relatedness between actors and recipients. The actual degree of relatedness between elephant allomothers and the young they care for is especially difficult to determine due to their long life spans, long interbirth intervals, and the dearth of long-term individual am histories. However, tentative assumptions regarding relatedness can be inferred horn past studies of elephant population biology (Laws and Parker, 1968; Hendrichs, 1971; Douglas-Hamilton and Douglas-Hamilton, 1975; Laws et al., 1975; Hanks, 1979). Full sibling relationships are unlikely to exist in allomother calf dyads unless specific males and females meet and breed times over extended interbirth intervals. Such long-term associations between individual rules and females have not been documented in the African elephant and seem unlikely, given the characteristic sexual segregation of elephant society and the frequently extensive dispersal of males. Another, more likely mating pattern would produce a genetically lower form of relatedness that has been overlooked in most considerations of allomothering among elephants. Sibships between allomothers may be based upon shared paternity.

Dominant bulls may obtain the majority of copulations with estrous females after initial contests between males, (Buss and Smith, 1966). Because females breed synchronously, single dominant bull may inseminate several females in the same herd. Female calves born to these herd mates form a cohort relationship. Such relationships may, in fact, form the tightest ties in elephant-kinship groups (Douglas -Hamilton and Douglas-Hamilton,1975). Over the years, the closeness of these peer associations may promote cooperation and reciprocity. In promiscuous species where single-offspring litters are the rule, each female produces a series of half-siblings. A third potential genetic basis for allomothering is that of maternal sibship.

Several characteristics of elephant family units suggest that this is the most likely form of relatedness between allomothers and the young for whom they provide care. Intercalf intervals have been estimated from demographic data on several elephant populations (Laws et al., 1975; Smuts, 1975; Hanks, 1969, 1979), and consistent patterns of age, intercalf intervals and placental scars strongly suggest that calves and allomothers frequently share the, same mother. Furthermore, most allomothers are significantly younger than the calf's true mother. If allomother-calf dyads are Indeed predominantly half-sibships, then by enhancing the probability of survival for their half-siblings, allomothers could profit by increasing their inclusive fitness. Just as the likelihood of allomothering may vary with the degree of relatedness, it may also vary according to the degree of need (Hamilton, 1964).

By providing help at critical times, allomothers might realize large gains in inclusive fitness with a relatively small investment of energy. Periods of illness, injury, malnutrition, or other stress on mother or calf would present these opportunities, as would difficult periods during calf development, such as weaning. Cows and calves experience conflict during weaning, like most: mammalian mother–young pairs (Trivers, 1974). Mothers frequently resist nursing attempts and often leave newly weaned calves behind when the herd moves on. If allomothers invest their energy Prudently, their effort should intensify during weaning, thereby aiding the dependent calf and freeing the mother to cycle and conceive again. Support for the hypothesis of differential care with respect to calf development was obtained from several herds at Tsavo West National Park, Kenya (Dublin, unpublished data, 1974, 1976). Table 3.1 shows that calves in the final stages of weaning, between the ages of 2 and 4 years, Interacted Significantly more ($P<.01$) with allomothers than those under the age of 2 or over the age of 4 years ($p<.05$).

Allomother-calf interactions vary as a function of the discrepancy in age between them. Where female helpers are close in age or size to smaller calves in the group, allomothering activities often involve play. Play bouts may help educate youngsters in use of their trunks for dusting, mud-bathing, showering, drinking, and feeding. Appropriate responses to external threats may also be taught. Play may also strengthen intraindividual bonds that facilitate cooperation later in life. When the allomother is substantially older than the calf, her behaviour is more maternal. Allomothers nurse, bathe, and direct daily movements of the young in these case. Age-asymmetric associations have been considered and examined more thoroughly than peer-like relationships in studies of elephant social behaviour by Douglas-Hamilton and Douglas-Hamilton (1975) and Laws et al. (1975).

Table 3.1. Dyadic Interactions between Calves and Allomothers

Age class[a] (months)	*No. of focal calves[b]*	*No. of dyadic interactions per observation day[c]*	
0-24	19	5.3	**
25-48	18	36.8	
49+	20	19.7	*

[a]Calves ages were estimated using methods developed by Sikes (1966,1968) and Douglas-Hamilton (1973).

[b]Focal sampling by the author of individual calves in Tsavo, West National Park, 1974-1976

[c]One observation day was 12 hours of continuous Observation, from 0600 until 1800.

*P<.05

**P<.01

It difficult to determine whether the aid given by allomothers helps to increase their inclusive fitness. This question will become more tractable, however, as data accumulate on survival and Its correlation to relatedness.

(b) Reciprocation

Long-term associations or bonds between related or unrelated individuals are hypothesized to increase the probability that cooperative acts will be reciprocated (Wrangham, 1980; Axelrod and Hamilton, 1981). In elephants, herd females are in contact for many years. Hence, a female who allomothers may be reciprocated for her efforts, either in the present or the future. Immediate reciprocation might-include cooperation in defense of her own offspring or aid in locating and acquiring critical resources. Future reciprocation would involve care offered her own offspring, either from other herd females or perhaps from calves she has allomothered. The remaining question for reciprocity is whether the allomothers is increased through these cooperative acts beyond that of herd females who do not allomother. Data presented in Section, II, C, 1, d indicate that allomothers are treated differently by mothers when they are with or without the calf. The information needed to resolve this questions would be a long-term tally of cooperative acts between individuals from which the degree of reciprocity might he measured directly.

(c) Experience

Young females also may profit through caring for the young of another cow if the experience increases the likelihood of successfully rearing their own offspring (Spencer-Booth, 1970; Lancaster, 1972). The cost to allomothered infants has been examined in Japanese monkeys (Kurland, 1977), langurs (Hrdy, 1976,1979), and baboons (Wasser, this volume) and found to be high in some cam. Although primate infants may suffer during allomothering from rough handling, separation from the mother, or decreased access to high quality forage or milk, elephant calves do not. Whereas allomothers could gain experience at the expense of another female's fitness in primates, such exploitative strategic are less likely in elephants because elephant allomothers are not permitted ready access to extremely young calves. Costs of allomothering are discussed in Section II, C, 2.

(d) Resource Access

Perhaps the most immediate benefit to allomothers is their unproved access to resources. Frequently, subdominant females are denied access to certain preferred foods or scarce resources (Douglas-Hamilton and Douglas-Hamilton, 1975; Dublin, unpublished data, 1974-1976). When in the company of a dominant female's young, however, these same sub-dominants are allowed greater access to water, mineral licks, and favoured foods., When not accompanied by a dominants calf, subdominant allomothers are often forcibly excluded from resources (Table 3.2). Elephants live mi environments that are periodically resource-limited. If maintaining proximity to dominants calves provides substantial benefits to, allomothers in the manner just described, one would expect a strong inverse correlation allornothering frequency, and habitat quality.

Table 3.2. Access to Valuable Resources for Subdominant Females with and without the Accompaniment of a Dominant's Calf

	Dominant female		
Subdominant female	*Allow access*	*Prevents access*	*Total attempts*
With calf	110	18	118
Without calf	9	75	84
Total responses	119	93	212
		$X^2 = 121.3$, $p<.001$	

2. Potential Costs of Allomothering

Allomothering may be costly as well as beneficial to the allomother. Allomothering activities consume time and energy. Directing or restraining the movements of young calves and helping them to procure food adds an additional burden to the self-maintenance energy budgets of adolescent cows, as it does for helpers in any species (Hamilton, 1964; Emlen, 1978). Thus, nulliparous or nonlactating females caring for other females young may have insufficient physical reserves remaining to produce their own offspring in a given year. Consequently, their reproduction may be delayed for reasons other than sexual immaturity alone. Impaired physical condition may result in delayed maturation, failure to ovulate, increased failure of implantation, or heavy *in utero* or postpartum mortality.

Poor body condition is known to cause deferred reproduction in mammalian species of many different orders (Sadleir, 1969) including ungulates (Preobrazhenskii, 1961; Wiltbank et al., 1962; Klein, 1970; Lamond 1970; Geist, 1971; Sinclair, 1977; Belonje and von Niekirk, 1975) and humans And McArthur, 1974, Pond, 1978). Lactating subdominants in the herd will allow the calves of older females to suckle (Douglas-Hamilton and Douglas-Hamilton, 1975; Kingdon, 1979, Bryceson 1980, Leuthold, 1977). And unconfirmed observations of calves being nursed by nulliparous females have been reported m captive and wild elephants. This suckling may result in lactational anestrus and thus reduce the wet nurse's reproductive capabilities during that time Kingdon, 1979).

III. Competition

In opposition to the proposed benefits of cooperative interactions a among group members, females may also sustain costs from intrasexual competition. Past emphasis on male-male competition in many socio-biological-analyses may have diverted attention from the form and function of competition among females, which by nature may be less overt. The most important ramifications of female-female competition (which occurs in various forms) are those that serve to increase the variance in reproductive success of individuals Kleiman, 1979). Over a female elephant's extensive reproductive lifetime, her competitive interactions will vary as a function of her age, reproductive state, and social status. Reproductive hierarchies may govern the ways by which variable, reproductive success is mediated.

A. The Importance of Timing to Reproductive Success

Successful reproduction in elephants requires adequate food, water, and shade during pregnancy and lactation (Douglas-Hamilton and Douglas-Hamilton, 1975). The timing of breeding is critical for ensuring that females can enhance their own condition prior to the time of conception and bear their young at the most propitious time of year (Williamson, 1976). Females in poor nutritional condition often do not conceive. Young in the height of the rains are threatened by exposure to severe weather conditions. Those born too late in the dry season face shortages of high quality forage (Laws et al, 1975). As a result, most elephant populations, show some degree of seasonality around the rainy period (Laws et al., 1975; Hanks 1979). Areas with bimodal rainfall distribution experience bimodal peaks of breeding activity (Laws et al., 1975; Kingdon, 1979).

Peak breeding activity occurs 1 or 2 months after the onset of the rains. This ensures that 22 months later births will occur either during the short rains or during the onset of the long rains (Hanks, 1969). Intrasexual competition for access to breeding opportunities may be expected during these optimal times.

B. Social Status and Breeding Opportunities

Females vary in their relative abilities to acquire the necessary resources for successful reproduction. These resources may include food, water, and mates. Competitive superiority In female elephants appears to be directly related to the individuals status in the breeding hierarchy. The underlying determinants of individual status among elephants are not well understood, but age, size, maternal rank, and the individuals disposition are important (Sikes, 1971; Kingdon, 1979). Large, old cows tend to be dominant individuals, and the largest is usually the matriarch. In this system, young subdominant females are likely to be the poorer competitors and therefore experience lower reproductive success. Within a given birth season, distinct separation of parturition appears between dominant and subdominant females. In my 1974-1976 Tsavo West study (Dublin, unpublished data), subdominant females gave birth later, on average, than older, dominant females.

During the long rains, calves born after the peak in mid-March incurred higher mortality than those born before the peak. Of 15 born after the rainfall peak, 8 (53%) died compared to 5 out of 23 (22%) of those born before the peak ($X^2 = 4.03$, $p < .05$). In contrast, during the short rains there was no significant difference in mortality between prepeak and postpeak births. Both groups experienced very low mortality. Given the unpredictability of resources over the long gestation and preweaning period of elephants, subdominant females may experience reproductive failure if they cannot breed during peak rainfalls. Where high population densities and/or scarce

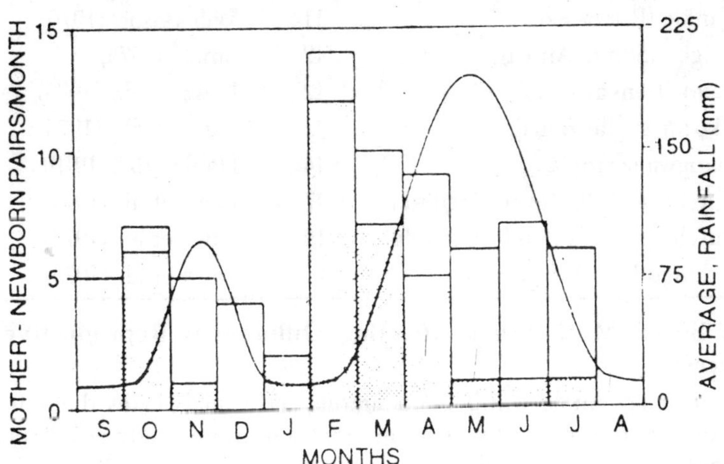

Fig. 3.3. Distribution of births to dominant (shaded bars) and subdominant (white bars) females in Tsavo West National Park herds in relationship to average monthly rainfall. Social status was determined from dyadic interactions.

resources compound this underlying unpredictability, natural selection should favour those females that defer their reproductive efforts until circumstances improve.

As resources become more abundant or females gain higher status in the hierarchy, they may acquire greater control over the timing of their own reproduction. This may help explain interpopulational variation in ages at first ovulation (Table 3.3). In many vertebrate species, young females confronted with high intrasexual competition merely disperse from the maternal home range. However, emigration horn the maternal herd is viable option for young female elephants. In a social system in which survival depends heavily on cooperative group efforts for protection and location of food, young animals are probably unable to survive on their own, or alternatively, to gain access to other herds. Apparently, intergroup segregation is maintained by aggression, which prevents unfamiliar animals from readily joining new groups (C. Moss, personal communication, 1981).

Table 3.3. Average Age at First Ovulation in Free ranging African Elephant Populations

Population (Country)	*Age (years)*	*Source*
Murchison Falls (Uganda)	7	Buss and smith (1966)
Wankie (Rhodesia)	11	Williamson (1976)
Kruger (South Africa)	12	Smuts (1975)
Tsavo (Kenya)	12	Laws et al., (1975)
Mkosmasi (Tanzania)	13	Laws et al., (1975)
Luangwa (Zambia)	14	Hanks (1972,1979)
Murchison Falls North (Uganda)	17	Laws et al. (1975)
Murchison Falls South (Uganda)	19	Laws et al. (1975)
Budongo Forest (Uganda)	23-25	Laws et al. (1975)

C. Social Mechanisms Affecting Differential Reproductive Success

Female-female competition among individuals takes different forms, expressed overtly at some times and more subtly at others. Kleiman (1980) reviewed the possible methods a dominant female may employ to maintain her reproductive superiority and those that a sub dominant may adopt to prevent untimely investment in reproduction. In most cases, the exact physiological mechanism controlling reproductive delay or disruption is not apparent. However, delayed, suppressed, or infertile ovulation, failed implantations; and spontaneous abortions are all possibilities. Evidence from a number of mammalian species suggests that social interactions may mediate the inhibition of subordinate reproduction more often than was formerly realized (see also Wasser; Silk and Boyd, this volume). Inhibition in the presence of their mothers is experienced by juvenile female cactus mice (Skryia, 1978), microtine rodents (Zejda, 1961,1967; Batzli et al, 1977), common marmosets (Abbott and Hearn, 1978; Kleiman, 1979), and Mongolian gerbils (Payman and Swanson, 1980). Delay of juvenile reproduction in the presence of dominant females living in hierarchical breeding communities has been found in house mice (Drickamer, 1974a, 1979, 1980, Massey and Vandenbergh, 1980), prairie deer mice (Terman, 1969, 1980, Lombardi and Whitsett, 1980; Lloyd, 1980; Christina, 1970, 1980), rhesus monkeys Drickamer, 1974b, 1980), dwarf mongoose (Rood, 1980), European badgers (Ahnlund, 1980), wolves (Zimen, 1971; Altmann, 1974; Packard and Mech, 1980), patas monkeys (Rowell, 1978), glada baboons (Alvarez, 1973; Dunbar, 1979,1980, Dunbar and Dunbar., 1977), and wild dogs (Frame and Frame, 1976, Frame et al., 1979).

In resource-limited conditions, dominant female elephants may use their competitive superiority to suppress the reproductive efforts of subdominants (Sikes, 1971). Inhibiting the reproduction of subdominant individuals would protect the dominants‘ own reproductive investment by reducing competition for themselves and their subsequent newborn calves. Correlations, between social rank and reproductive demographics of elephant family units have not been established. However, physiological inhibition of reproduction may be triggered by social stresses imposed on subdominant females ‘by their mothers and other higher-ranking females (Sikes, 1971). In addition to physiological mechanisms, dominants may also limit subdominants access to fertile mates, thereby further reducing reproductive opportunity. Potential methods of reproductive inhibition are discussed in the following paragraphs.

1. Delaying or Inhibiting Ovulation

If stress is a significant factor in delaying or inhibiting estrous cycles, dominants may effectively influence the reproductive efforts of subdominant Individuals through directed agonistic behaviours and other forms of social, harassment. Estrous cycles may subsequently fall out of synchrony with the optimal. Breeding periods discussed in Sections III A and III B. If estrus in the African elephant resembles that reported for the Asian elephant, *Elephas maximus*, fertile periods are very short and are separated by period of infertility lasting up to 4 months (Schmidt, personal communication, 1981). Any delay or Interruption of the ovulatory cycle may a cow to cycle one or several more times before successful, conception.

The interovulation period may be shorter in the African elephant (Short, 1966). Nonetheless, delay could still result in reproductive readiness (during suboptimal periods. For example, by the time a cow recycles, cow-cow-calf herds may have already returned by dry sewn ranges far from breeding males. Also, decreased habitat quality late in the season would lower the physical condition of females and further delay estrus (Laws and Parker, 1968; Sadleir, 1969). Additionally, if conception occurs in the dry season, with parturition following 22 months later, claves would be born under extremely poor conditions, and increased mortality would be expected. (see Section III, A and III, B) (I Laws et al., 1975; Guinness, Albon and Clutton-Brock, 1978; Guinness, Clutton-Brock, and Albon, 1978).

2. Social Aggression

Seasonal changes observed in female-female interactions support the prediction that dominant individuals influence the reproductive efforts of subdominants through agonistic behaviour. In four Tsavo West herds, dominant females significantly decreased their agonistic encounters with subdominant females, but significantly decreased their agonism towards adolescent males during the peak breeding period (Table 3.4). Subdominant females may represent a direct competitive threat for mate access (see Section III, C, 4) and optimally timed breeding opportunities, whereas young bulk compete only for food. Herd males, due to their relatively higher energetic demands, may be a greater threat to herd matriarchs and their young outside the breeding season when they vie for access to food, water, and mineral deposits. If stress can significantly reduce reproductive success, increased aggression could delay or, inhibit fertile ovulations. Circumstantial evidence supports this hypothesis (we Section III, C 3).

Table 3.4. Differential Aggression of Dominant Females toward Male and Female Subdominants between Prepeak and Peak Breeding Periods

	Subdominant females			Subdominant males		
Herd	*n*	*Trend in aggression*	*P- value*	*n*	*Trend in aggression*	*P- value*[a]
A	5	Increase	<.01	5	Decrease	<.05
B	4	Increase	<.05	3	Decrease	<.01
C	7	Increase	<.001	6	No change	<.10
D	3	No change	<.10	6	Decrease	<.05

[a]Data were analyzed using a paired t-test on observations of known individuals before and during the peak breeding period.

3. Inducing Infertile Ovulations

In 350 B.C., Aristotle noted physiological evidence that first suggested that infertile ovulations occur in elephants. Sikes (1971) noted that adult female elephants and nulliparous or young cows frequently possess multiple accumulated corpora lutea (Sikes, 1971). These corpora lutea vary in size, shape, and numbers (Buss and Smith, 1966). Up to 26 have been observed in a single female (Hanks, 1979). Short (1972); Smith et al. (1969); and Hanks (1979) speculated that a 'critical mass' of luteal tissue may be required before conception is possible. Fully formed, these corpora lutea have a degenerate appearance and show little or no hormonal activity.

Progesterone, the hormone produced by corpora lutea-for maintaining pregnancy, is present in only minute amounts relative to that of other mammalian species studied (Short, 1972, Hanks, 1979). If the abundance or size of corpora lutea masses in individuals is negatively correlated with status in the dominance hiearchy, corpora lutea accumulations my represent an infertile state resulting from social inhibition, as hypothesized by Sikes (1971). These data, however, are presently unavailable. Herds in marginal habitats or dense populations would be most likely to show the correlation if intrasexual competition is under resource limitations.

4. Limiting Access to Mates

Opportunities for subdominant females to mate successfully in a given breeding season may also be controlled by dominant females. These females lead their herds from dry season home ranges to prime rainy season sites, where the majority of matings take place (Laws et al., 1975; Kingdon, 1979). Due to their size and social status, dominant females have access to the best forage on the breeding grounds. Consequently, these females are In the best physical condition. They also are the most experienced breeders. Reproductively dominant females thus enter estrus earlier (smuts, 1975; Hanks, 1979) and breed first. The early breeding efforts of bulls may be monopolized by dominant females.

The reduced breeding opportunity of subdominant females may be exacerbated by a resulting temporary loss of libido in breeding males (Schmidt, personal communication, 1981) and perhaps reduced fertility following multiple copulations. Females may also compete amongst themselves through prolonged sexual receptivity (see Wasser, this volume). In captive populations female elephants will allow mounting and copulation even when unable to conceive. This includes copulations, during pregnancy, lactational anestrus, and anovulatory heats (Schmidt, personal communication, 1981). In this way, reproductively dominant females can complete, at little cost to themselves, by monopolizing the reproductive activities of eligible male. The occurrence of copulations outside of fertile periods has been reported in a spectrum of vertebrates including herons (Gladstone, 1979) primates (Kleiman and Mack, 1977, Hrdy, 1977, Rowell, 1978, Wasser, this volume), and lions (Bertram, 1973).

IV. Conclusions

Recent interest in the evolution of female social strategies has drawn attention to the matriarchal system of the African elephant. The relatively long and overlapping lifespans of individual elephants have provided an evolutionary basis for transmission of knowledge between generations and mediation of elaborate social controls between group members. Such advanced forms of social evolution have been most widely reported among primate species and have been attributed primarily to advanced brain development. However, the long lives of elephants, relative to those of other mammals, may have contributed as much a elephants' social development as intelligence alone. Elephant societies have traditionally been viewed as the epitome of female cooperation.

Emphasis has been placed on the overt cooperative acts for which female elephants are well known -group defense and allornothering. Less attention has been directed toward competition among these same family members. Yet the importance of considering both cooperative and competitive interactions has become apparent. As Kingdon (1979) succinctly stated with regard to elephants," Interdependence and mutual support under some conditions are not necessarily incompatible with members being competitive in other situations." For elephants, selective pressures for transmission of knowledge about spatio-temporal resource distribution coupled with predator defense, lend weight: to the adaptive value of group cooperation.

On the other hand, competitive interactions are also likely In a species living under periodic resource constraints. In a dominance-ranked social system in which sociality is essential, competition will be felt most severely by young or subdominant females. Evidence suggests that this competition may have its greatest cost to low-ranking females through social inhibition of their reproductive efforts. The social interactions that lead to suppressed, delayed, or disrupted reproductive attempts can be obvious or subtle but may be the most potent form of intrasexual competition among females. Among elephants, where individual females may be associated for decades herd integrity clearly requires a delicate balance of cooperation.

4

Evolution of Social Behaviour

Most mammals evolved sociality via the familiar route. In simpler systems, solitary females or pairs care for their young. As the period of parental care increases, earlier young are still present when the next generation is born and they often help rear their younger siblings. In most mammals, female offspring remain in their natal groups (the group in which they were born) but males tend to leave, or are driven out. After a period of wandering, males insert themselves into some other social group. The parasocial route has been taken by a few bats, in which stable colonies of unrelated females defend Young jackals often remain with their parents until the next litter is produced. The older offspring help their parents find food for and defend their younger siblings. In the photo, an older male offspring who is helping his parents is being groomed by the mated pair. After a time, however, the older offspring, especially the males, are driven away from the family group and must insert themselves into another social group–or they must found a new group. Like many other diurnal primates, baboons live in groups and exhibit many different social interactions. Members of a troop both compete and cooperate with one another. In this picture, a male is harassing another male in a dug-out depression (head barely visible), while nearby there is a consorting pair in which a male is grooming a female. Males compete for access to females, yet they often groom one another. Whether two particular individuals are engaged in competitive or cooperation activities may change many times each day.

Communal Foraging Areas

In a few of these species, large numbers of females place their young in a central creche, or nursery. Each female gives birth to a single young, but females often suckle more than one infant, suggesting that at least some communal care of offspring has evolved. Primates exhibit the full range of mammalian social systems. Nocturnal forest-dwelling insectivores, such as lorises, some lemurs, and the owl monkey, live in pairs and usually forage solitarily. May diurnal species take insects and other animal food when they are available, but most of them depend primarily upon fruits, seeds, and leaves for their energy. Some of these species, such as lemurs and gibbons, live in very small groups consisting of a single pair and their offspring.

Males and females are nearly the same size, as is the case in the nocturnal species. Some of the most familiar number of reproductively active males and females and their offspring. Groups tend to be largest in species that feed mostly on the ground and smallest in arboreal species. Males are larger than females in these multimate groups and they play a major role in defense of the troop. The sizes of the troops vary with environmental productivity, being larger in food-rich environments than in poorer ones. Yellow and olive baboons live in moist, productive savannahs. Their troops contain up to several dozen or more individuals. The hamadryas baboon lives in very and areas where each troop must cover a wide range to find food. Its groups are small, each with a single mature male. As with other mammals, female primates usually remain with their natal troop, whereas young males are driven away by older males before they become reproductively active.

In multimale troops, there are strong dominance hierarchies among males, and a few of the males do most of the copulating. Females do not form marked dominance relationships, but in some species, such as vervets, they interact a great deal, frequently grooming one another. In other species, such as chimpanzees, females have little to do with one another. Human social systems are more similar to those of social carnivores than to the systems of other primates. This is because male carnivores can make more significant contributions than male herbivores to the care and feeding of their offspring.

The development of cooperative hunting of large mammals by our ancestors led to the formation of cohesive all-male hunting

Fig. 4.1. The pig-tailed macaque of S.E. Asia (Macaca) and a South American capuchin monkey (Cebus).

groups, heightened intragroup cooperation, and division of labor among the sexes. The quantity of resources controlled by a man strongly influences his access to the reproductive potential of women. This is especially marked in societies where marriages are arranged and parents do not release their daughters until a man can deliver sufficient resources to them (bride price). The control of most resources by men reverses the typical mammalian pattern of sexual mobility.

In most human societies, men remain with their natal groups and women transfer to the groups of their mates. Nonetheless, like those of other primates, human societies are built around kinship groups. All human societies have elaborate systems for identifying kin and rules for sharing resources among relatives. These patterns suggest that human societies evolved via the familial route, supplemented by cooperation among closely related adults of the same generation.

Chemical Communication

Most mammals have a highly-developed sense of smell; the olfactory sense organs are borne on a group of scroll bones termed the endoturbinales, and sensitivity to odours relates to the number of receptors, which can be approximated from the surface area of these bones. Their area in the German Shepherd dog is 200 cm^2 as compared with 4–5 cm^2 in a human being; this disparity is reflected in the fact that a dog can detect acetic acid molecules at a concentration 100 million times less than the minimum which can be perceived by the human nose. Some mammals possess an independently innervated additional olfactory organ known as the vomero-nasal or Jacobson's organ, which lies anteriorly on the roof of the mouth and may connect with the nasal cavity. Lip curling or 'flehmen' is associated with the possession of this organ and is seen in many mammals during courtship (e.g., stallions on scenting the urine of an oestrous mare). It occurs in marsupials and several orders of placental mammals. Correlated with the well-developed sense of smell in most mammals is a chemical communication system which involves the possession of special odour-producing glands.

The actual odours vary from species to species; in some cases the scent results from the secretion of a single chemical such as musk (from the preputial glands of the musk deer, *Moschus moschiferus*), *civet* (from the anal glands of the civet, *Civettictis civetta*) and *castoreum* (from the preputial glands of the beaver, Castor fiber). In other cases, the message is conveyed by a mixture of chemicals, as in the vaginal secretion of oestrous rhesus monkeys (*Macaca mulatta*), lion (*Panthera leo*) and red foxes, where the odour is produced by bacterial decay of secretions. This can be demonstrated

Fig. 4.2. The Chimpanzee (Pan).

by treating the vagina of the female rhesus macaque with antibiotic, which renders her sexually unattractive to males. In many species urine may contain odorous secretions which convey messages to conspecifics. This is a common method by which female mammals signal that they are in oestrous. The odours produced by mammals may communicate to conspecifics in either of two ways.

Firstly, they may have a priming effect, whereby they influence the physiological state of the recipient with a delayed behavioural effect. An example of priming odour is the Bruce effect. If a female mouse is mated and conceived but, within five days of mating, is exposed for a minimum of twelve hours to the scent of an unfamiliar male, her pregnancy terminates and she comes back into oestrus. Secondly, odours may have a direct influence on the behaviour of a conspecific, among the best known being sexual attractants secreted by the female. The scents of mammals are normally produced by dermal glands of two kinds. Apocrine or sweat glands secrete droplets of aqueous fluid and small pieces of cell debris. They may either be widely spread over the body, as in horses and man, or concentrated to form glandular areas such as the chin glands of rabbits and the dorsal glands of peccaries (*Tayassu sp.*). The second types of gland is the holocrine or sebaceous gland, which produces a secretion known as sebum.

Holocrine glands may be generally distributed about the body, as in the rat, or concentrated as in the ventral glands of the mongolian gerbil (*Meriones unguiculatus*), the anal glands of many carnivores and the preputial glands of rodents. Artiodactyla have well-developed preorbital, tarsal, metatarsal or interdigital holocrine glands. Some glands are under hormonal control, so that the scent of the female may change throughout the oestrous cycle. Male and female mammals may differ in their sensitivity to various chemicals; this is true even in our own species where women are more sensitive to musk-like substances than are men. *Le Magnen* tested the sensitivity of men and women to an artificial musk, exaltolide, and found that 40–50% of men could not perceive this substance at all. The average sensitivity of those men who could detect it was 1000 times less than that of women.

There is some evidence for a link with oestrogen production, for ovariectomized women's sensitivity was found to be 100–1000 times reduced, but was restored if they were treated with oestrogen. *Le Magnen* also found that women's sensitivity to musks varied

Fig. 4.3. Gorilla.

during the oestrogen were at a peak. The significance of these findings for human beings is not clear but they shown that the mammalian reproductive cycle can influence both the secretion of odours and an individual's sensitivity to them. In discussing mammalian odours it seems best to avoid the use of the term 'pheromone' which is commonly applied to some insect odours, as few mammalian odours elicit the stereotyped responses characteristic of pheromones. A number of techniques have been used to determine a mammal's response to, and capacity to distinguish between, odours. Examples are: direct observation of an individual exposed to an odour, comparison with a control whose sense of smell has been eliminated (anosmic control) or discrimination experiments in which, either the animal's natural choice is recorded, or one odour has come to be associated with a reward (positive reinforcement).

In some instances odours have been analysed by gas chromatogram, their active principles identified and the animal's response to synthetic odours tested. Mammalian chemical communication occurs in a number of different behavioural contexts, which will now be considered in more detail.

The Use of Odours in Social Communication

Discrimination tests using an odour from an individual's own species and that of a closely related species have shown that many mammals can recognize their species by scent, for example, guinea pigs, Mongolian gerbils, and brown lemurs (*Lemur fulvus*). The ability to discriminate, and show a preference for, the odour of their own subspecies has been demonstrated experimentally in bank voles (*Clethrionomys glareolus*) blacktailed and mule *deer (Odocoileus hemionus*) and saddlebacked tamarins (*Saguinus fuscicollis*). An extremely important piece of information which may be conveyed by scent is the sex of an individual, and male pigs (*Sus scrofa*), guinea pigs (*Cavia porcellus*), Mongolian gerbils, brown rats and house mice can distinguish between urine from males and females. Similarly, brown lemurs and saddlebacked tamarins can discriminate between the scent marks of males and females. *Dixon* and *Mackintosh* showed that an odour in the urine of male mice influences the aggressive behaviour of other males. They rubbed castrated male laboratory mice with either male or female urine and found that the former elicited more aggression from an intact male opponent. Urine from castrated males and ovariectomized females did not

Fig. 4.4. Cervulus muntjac (Barking deer) male.

produce this effect. The influence of olfactory cues on sexual behaviour has been reviewed by *Keverne.* Some male mammals in rut produce odours which attract females.

This is true of the European hedgehog (*Erinaceus europaeus*), where a secretion produced by the penis has a stimulating effect on the courtship behaviour of the female; the rutting male's urine is also attractive. The odour of the male domesticated pig has a specific behavioural effect on the female in inducing the stance which she adopts prior to copulation. Male Asiatic elephants (*Elephas maximus*) produce a copious secretion from the temporal gland when in 'musth', a state in which they become exceptionally aggressive and sexually active for short periods of time. Certain odours communicate the state of maturity of an individual. The urine of juvenile mice has a distinctive smell which reduces the aggressiveness of older individuals towards them, and the odours produced by the scent glands of European rabbits (*Oryctolagus cuniclus*) are less intense in juveniles, so that even human beings can roughly age young rabbits on the basis of odour alone.

Adult mammals frequently recognize their own young by scent; in the case of the domesticated goat (*Capra hircus*), *Gubernick* has convincingly demonstrated that the mother labels her newborn kid by licking it and allowing it to take milk, which imparts the mother's own smell to her offspring and allows her to recognize it subsequently. Kids which have not been licked or suckled are rejected by the mother goat. In the case of tree shrews (*Tupaia glis*) and dwarf mongooses (*Helogale parvula*) both parents mark the young with their scent glands. Generally female mammals are only fertile for brief periods between pregnancies or, in seasonally breeding species, at a certain time of the year.

The fertile period itself may last only two or three days and most females indicate their sexual receptiveness to the male by means of a vaginal secretion. *Michael* and *Keverne* and *Keverne* showed that in the rhesus monkey the secretion is under the control of oestrogen, while in golden hamsters (*Mesocricetus auratus*) the male's preference for the vaginal secretion of a oestrous female is remove by castration but restored by treatment with testosterone. Female sexual receptivity may also be indicated by an odour in the urine at oestrus. When urinating, the female is able to signal her sexual receptiveness to the male even when she is not physically present, an ability which must be extremely important for solitary mammals.

Male dogs, rabbits, mice and horses have all been found to be attracted to the urine of oestrous females; this phenomenon is probable widespread among mammals, but remains to be investigated in other species. Social mammals behave differently towards strangers and members of their own colony, and, in some species, colony odours are implicated.

In the case of the laboratory rat, an individual which is removed from its social group for a few days is attacked as if it were a stranger when reintroduced. If, however, the isolate is kept with soiled bedding from the colony cage, it is accepted back amicably. In such cases, the colony scent appears to be acquired from the general environment and there is no evidence for a specific colony odour. In other cases, however, the dominant member of a social group scent-marks other members of the colony, sometimes by rubbing its scent glands on the other individuals as in the case of the flying phalanger. Some hystricomorph rodents (*Cavia, Myoprocta, Octodon, Dolichotis and Cuniculus*) mark other colony members by spraying urine on to them; this is termed '*enurination*'. In desert rodents such as *Pediolagus*, sandbathing by all family members in the same spot may be a means of maintaining a group odour. Most mammals are capable of recognizing conspecifics as individuals. This faculty enables a mammal to recall its past social interactions with a particular individual and this to remember whether it was aggressive or friendly, dominant or subordinate. It is the ability to recognize individuals and recall their past behaviour which distinguishes the membership of a society from simply being part of an aggregation.

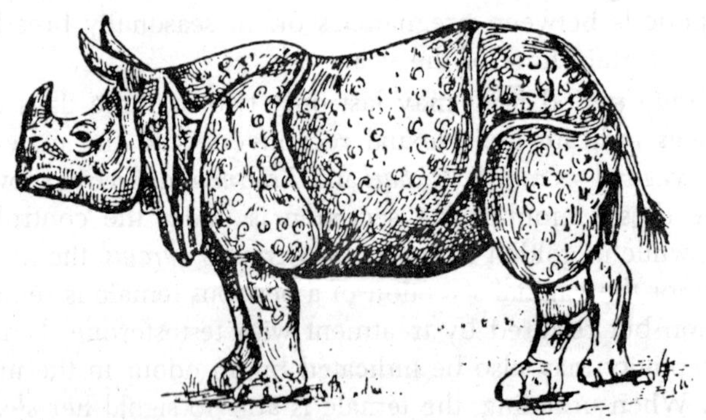

Fig. 4.5. Rhinoceros unicornis (Rhinoceros).

The ability to recognize other individuals by their odour is known to have developed in a number of mammalian orders which includes Marsupialia, Rodentia, Primates and Carnivora. The individual may not only be distinguishable by its scent at close quarters but may also leave individually recognizable scent marks throughout the area in which it lives, indicative of its personal occupancy of the area. Such behaviour is found in all mammalian orders with the exceptions of the aquatic Sirenia and Cetacea. The form taken by scent-marking varies from species to species; the male wolf urinates on strange conspecific urine from which *Fox* and *Cohen* deduced that the animal is essentially leaving a record of its presence as an individual.

Dwarf mongooses scent-mark, and *Rasa*, in an experimental study, found that these animals cannot only identify individuals by their scent marks, but are also capable of determining whether the scent was deposited recently. The black rhinoceros (*Diceros bicornis*) marks its territory both by spraying urine, which it disperses with rapid vibratory movements of the tail, and by leaving faeces at definite points in its territory. Faeces are also used for making the European badgers (*Meles meles*), weasels (*Mustela nivalis*), otters (*Lutra lutra*) and giant rats (*Cricetomys gambianus*). Hippopotamuses (*Hippopotamus amphibius*) use lateral movements of the tail to spread faeces over considerable distances. A number of mammals mark the environment directly with their scent glands. Some species (mustelids, bush dogs, and marmosets) use the anal glands, others use preorbital glands (ruminant artiodactyls), sternal glands (Mongolian gerbils) or chin glands (rabbits).

Fig. 4.6. Hippopotamus amphibius (Hippopotamus).

Discrimination experiments have demonstrated that individual recognition is possible from scent marks in ring-tailed lemurs (*Lemur catta*) and dwarf mongooses. This does not necessarily mean that the animal uses scent marks in this way, under natural conditions, but as a general inference it seems reasonable. More field and experimental work would undoubtedly be of immense value. One piece of information which may be carried by an odour is the status of an individual. To take the simplest case, a dominant may scent-mark more frequently than a subordinate, and also deposit the chemical signal more frequently both on conspicuous objects in the environment and the females of his harem. Laboratory rats and mice are capable of discriminating between the body odours of high and low-ranking males, females showing an active preference for the scent of the dominant.

It seems highly probable that these differences are related to hormonal states, as subordinate mice have heavier adrenal glands and lower plasma testosterone levels. Some mammals produce an odour when alarmed, and those of the woodchuck (*Marmota monax*) and the black-tailed deer (*Odocoileus hemionus columbianus*) cause active avoidance by conspecifics. In some mustelids such as the European polecat (*Mustela putorius*), striped skunk (*mephitis mephitis*) and wolverine (*Gulo gulo*), the anal glands secrete a foul-smelling fluid; in the case of the skunk this; is sprayed at the predator, the active ingredients in this species being trans-2 butane-1-thiol, 3-methyl-1-butanethiol and trans-2-butenyl methyl disulphide. The extent to which these scents cause alarm to conspecifics is not known but, as the species are rather solitary, this aspect is unlikely to be important.

In this case glands which are normally used for social communication in mammals have become modified to produce an aversive effect on members of other species. A number of functions have been ascribed to scent-marking in the animal's home range. *Eisenberg* and *Kleiman* suggested that scent marks help an individual to orientate in its own *PAGE 139* living area and that a second function is to advertise to conspecifics the presence of an individual or park in a particular region. Generally, the presence of a scant mark does not, of itself, appear to act as a deterrent to trespass, although lone wolves have been shown to avoid areas recently marked by packs, and cheetahs change direction on encountering a fresh scent mark. However, *Wells* and *Bekoff* found that coyotes trespass into areas which have been scent-marked by other individuals, making it likely that the odour in this case acts only to advertise the presence, and possibly strength of numbers, of conspecifics.

The individual's past experience may tell it whom to avoid or seek out, depending upon its established relationship with the scent-marker. *Brown* divided social chemical signals into two kinds. The first type are identifier odours which are stable for periods of time, for example individual, colony, sex-specific, age-specific, and species-typical odours. The second type are emotive odours which are released in special circumstances and are not long-lasting, for example rut, social status, maternal and alarm odours. It must also be borne in mind that the *directness* of the social effect varies; signal odours produce an immediate effect, while primer odours have longer-term physiological effects. Clearly, most mammals which rely heavily on chemical communication are capable of receiving a wide variety of messages by this means.

Eisenberg and *Kleiman*, in their review of olfactory communication in mammals, summarize the probable role of odours by stating that–for each individual depending upon his age, sex, mood and reproductive state there is an optimum odour field which will provide the optimum level of security. This odour field is composed of a combination of olfactory stimuli emanating from the individual, the environment, the conspecifics. If a disturbance in the odour filed occurs (a) through a change in physiological state which alters the sensitivity of the individual, and his perception of the optimum odour field (b) through the introduction of a foreign odour, or (c) through a change in the odour field caused by the dissipation of scent previously deposited, then the individual will attempt to restore the previous balance by the release and deposition of scent. The changed odour field may also arouse the individual, depending upon the nature of the change, and other behaviour may be initiated.

Acoustic Communication

Sound Production

Most mammals produce sounds by means of the respiratory tract, using inspired or expired air. Sounds are usually formed in the larynx by the vocal cords and then modified by changes in shape of the buccal cavity. Some species, such as the siamang (*Symphalangus syndactylus*) and organ utan (*Pongo pygmaeus*) possess special elastic-walled vocal sacs which resonate and serve to amplify the sound and reduce harmonics in the call, while South American howler monkeys (*Aloutta sp.*) have a specialized inflated hyoid bone which acts as a call amplifier; such adaptations enable the sender to be audible over distances of several kilometres. Some gazelles can produce sounds in their noses by vibrating cartilaginous structures or folds of skin, and cetaceans use the blowhole and nasal passages to vocalize.

Fig. 4.7. Pongo satyrus.

Some mammals, however, produce sounds from structures other than the respiratory tract; many rodents grind their teeth when aggressive, mountain gorillas (*Gorilla gorilla berengei*) beat their chests while chimpanzees (*Pan troglodytes*) drum with their hands on hollow trees, and black-tailed prairie dogs (*Cynomys ludovicianus*) stamp on the ground. Species whose hair is modified to form spines, such as the Bornean rattle porcupine (*Hystrix crassispinis*) and the tenrecs (*Centetes* and *Hemicentetes*) can produce sounds by rattling their quills. There are certain physical constraints on sound production. Most mammals are capable of making high-pitched sounds, but larger mammals are also capable of producing low-pitched sounds because the length of the vibrating structure must be adequate to produce long wavelengths (compare for example a violin and double bass) and loudness is dependent upon the extent to which the vibrator can be displaced laterally (wave amplitude); thus shrews do not roar, but lions can!

It follows from this that one would expect that the lowest pitch obtained might change during the animal's development, and that a young animal would be expected to produce fewer low-pitched sounds that an adult. This is shown in Figure for a monkey, the crowned guenon (*Cercopithecus pogonias*), which has a cohesion call that functions to keep the group together. The data shown are for a male during prepuberty and puberty phases; the mean frequency for the low-pitched component and its standard deviation are given, together with the weight curve for the individual. The range of sounds produced by different species is very variable and is dependent upon the extent to which the animal also relies on other forms of communication and on its degree of sociability. For example, shrews, in which optical communication is virtually non-existent, rely heavily on acoustic messages, and the shorttail shrew (*Blarina brevicauda*) possesses a repertoire of at least five vocal signals.

Fig. 4.8. Meles (Badger).

Highly social species of mammal may have a wider repertoire of acoustic signals. Wolves, for example, have seven and chimpanzees twenty-four classes of vocal sounds. The kinds of signal produced are also dependent upon he environment, since this places physical constraints on the effectiveness of different forms of sound. In dense woodland, for example, long-distance calls tend to have a low pitch because high-pitched sounds are easily scattered and absorbed by foliage. Other factors which reduce the effectiveness of acoustic signals are wind, high humidity, and high temperature. *Marler* noted that diurnal monkeys which live in dense habitats tend to have discrete calls, whereas those living in open habitats tend to intergrade. Sound production may be different in the two sexes and in polygynous mammals such as red deer and howler monkeys the males alone produce the loud spacing call. In the great apes, *Fossey* showed that there is a considerable difference between the frequency of vocalizations by adult male chimpanzees and those of mountain gorillas; in the latter species, almost all of the vocalizations recorded were produced by adult males.

For its effectiveness, vocal communication depends upon the perception of the receiver, all mammals having binaural hearing which provides a mechanism for locating the source of a sound.

This is dealt with in the central nervous system which estimates the differences in the time, (of arrival) differences in intensity (amplitude) and differences in phase of the waves arriving at the two ears. The distance of the caller from the recipient of the signal may be estimated from the loudness (amplitude) of the call which decreases with the square of the distance. Another mechanism which may operate depends upon the fact that the rate of attenuation, i.e., decrease in loudness, is directly related to frequency. High-pitched sounds attenuate more rapidly than those of equivalent amplitude by lower frequency. These physical principles are important for interpreting the character of different forms of acoustic communication.

Table 4.1. Frequency of calling by different age/sex classes in chimpanzees and mountain gorillas

% of calls made by each class member	*Adult M*	*Adult F*	*Juvenile*	*Infant*
Gorilla	92	4	0.6	3
Chimpanzee	27	25	24	24

A social mammal faced with a predator may give a mobbing call which consists of loud, short notes converting a range of frequencies. This allows conspecifics to locate the signaller and hence the predator and, in some species, to go to the signaller's defence. Alternatively, an individual which discovers a predator may give a distress call which is usually a long call, high-pitched with a limited frequency range, providing little information as to the whereabouts of the caller. Intimate contact calls between members of a social group, by contrast, are usually intermittent, low-pitched and of relatively low amplitude, so that they can be easily located from nearby. Rhythmical variations or modulation of the frequency or amplitude of calls may also provide cues as to the whereabouts of the signaller.

The Messages Conveyed by Sounds

For convenience, vocal communication can be divided into two types–firstly, distant and secondly, intimate signalling. Alarm calls are a typical form of distance signal; they are loud are usually high-pitched, and there is often considerable similarity between the alarm calls of different species. These calls alert conspecifics to the presence of danger. Some species give different alarm calls when sighting different predators. Four such distinct vocalizations have been identified predators. Four such distinct vocalizations have been identified in the vervet monkey (*Cercopithecus aethiops*) in response to the following snakes small bird or mammal predators, large predatory birds and a major did or mammal predator close proximity to the caller. Belding's ground squirrels (*Spermophilus beldingi*), which are small rodents living in social groups in the Rocky Mountains, give two kinds of vocalization in response to predators, the first a chattering sound with a wide frequency range when sighting a ground predator such as a weasel, and the second a high-pitched whistle in response to an aerial predator.

Clearly the chattering draws attention to the caller's, and hence the predator's, precise location in the colony, whereas the whistle simply indicates the predator's presence, its actual position being irrelevant. This example may be used to discuss the theoretical problem associated with the evolution of alarm calls. Alarm calls present an interesting problem. An individual making up alarm call would appear to decrease its own fitness by drawing a predator's attention to itself, while at the same time increasing the fitness of other colony members. While a number of explanations have been put forward, ranging from group selection to the suggestion that the call may actually in some way deter the predator from attacking, the most plausible explanation is based on kin selection. For the alarm call to increase the individual's inclusive fitness, it must be assumed that the small extra risk incurred by the caller in drawing attention to itself will be offset by the consequent reduction of risk to a nearby related individual previously unaware of the danger.

A factor which is of utmost importance in this model is the likelihood of the caller's warning a close relative; if he has no relatives in the group, he would not be expected to give an alarm call which would serve only to increase the fitness of unrelated individuals. If on the other hand, the group contains a number of close relatives, the more likely an alarm call will be to save one of them and thus increase his own inclusive fitness. Data supporting this hypothesis have been provided by *Sherman* for Belding's ground squirrel (*S. beldingi*). He found that predators such as weasels and coyotes were often attracted to individuals giving alarm calls, so that there could be no doubt of the decrease in fitness to the caller. In this species, males migrate to breed whereas most females remain in their natal colony.

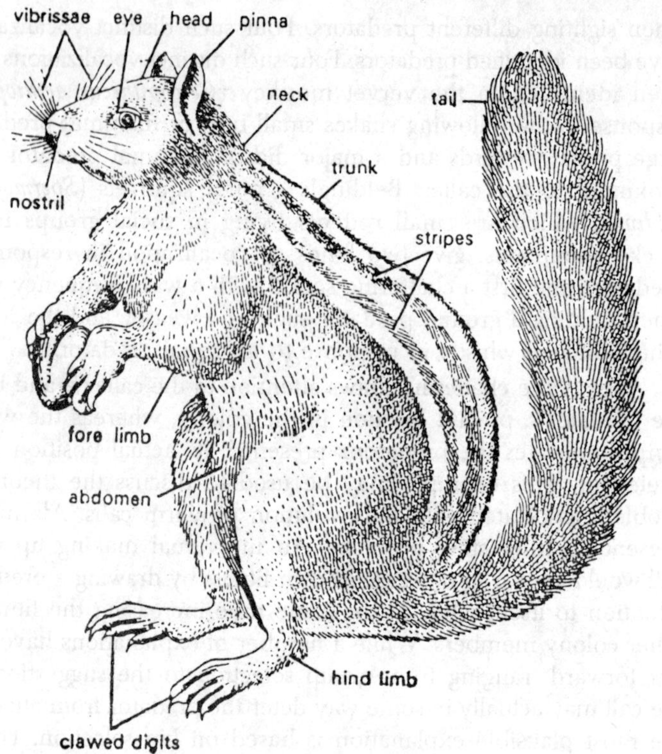

Fig. 4.9. Funambulus.

Kin selection would predict that females should give more alarm calls than males, and *Sherman* found this to be the case. That this was not simply a sex-linked trait was apparent from the fact that those females which had actually migrated into the colony gave fewer calls than those born into it. Another form of long-distance calling is that associated with territory ownership and spacing. Examples are the roaring of lions, howling of wolves, calling of prairie dogs, and the loud calls of the South American howler monkey, the African black and white colobus monkey (*Colobus guereza*) and the Asian gibbons (Hylobates) and siamang (Symphalangus). Two of the best-analysed examples will be chosen for more detailed consideration : the wolf (*Canis lupus*) and the lesser apes (Hylobatidae). The howling of wolf packs has been studied in detail by *Mech* and his co-workers.

Howling serves both to reassemble separated pack members and to space out different packs, i.e., it attracts familiar members and repels strangers. If, however, a howling stranger appears at close quarters, (within 200 metres), the alpha members of the pack may approach and investigate. *Harrington* and *Mech* studied radio-marked wolves from the air and also, by making simulated wolf howls, considered the effect of howling on packs and the frequency of howling in lone, as opposed to pack-living, wolves. *Lone* wolves howled much less than pack wolves and for a shorter period of time. Howling by the latter was particularly associated with kills, when the response of a nearby pack was to move away. As pack members howl together, it is possible that other wolves may from the howling be able to estimate pack size; in addition, each wolf has an individually identifiable howl. Generally pack howling lasts for about 40 minutes and is made up of with 85-second bouts of howling separated by 15-minute refractory periods. The wolf's howl is an extremely efficient method of territorial communication, which can be detected even by the human ear at a distance of eight kilometres and is audible over an area of 200 square kilometres. It provides information on identity of callers, numbers in the pack, and location; it also enables lost members to find their social group.

It is interesting that howling in the closely related coyote (*Canis latrans*), a much less sociable member of the dog family, acts only as a spacing mechanism and does not attract conspecifics. The songs of the Hylobatidae (gibbons and siamang) are a form of long-distance territorial calling which has been studied in some detail in both field and zoo conditions. Calling normally begins just before dawn, first one group and then another becoming involved. Choruses may be all-male, all-female or male-female duets, depending upon the species. The calls are distinct as regards individual identity and sex. In species which duet, the male and female sing alternately in synchrony so that the song sounds as though made by a single individual. It probably informs other gibbons of the mated status of the pair. Gibbons with territorial neighbours sing more than those with relatively isolated territories. Undoubtedly the most remarkable and complex of acoustic signals produced by mammals are the songs of the humpbacked whale (*Megaptera novaeangliae*).

Each song may last for as long as 30 minutes and can carry for very long distances. The song covers a great range of frequencies and each whale sings a unique song of remarkable consistency; there even appear to be dialects associated with different geographical regions. Thus the song of the hump-backed whale provides information on identity, area of origin and location; its function is, however, uncertain. Another interesting form of acoustic communication is the chimpanzee 'carnival' described by *Reynolds* and Reynolds where whole groups of individuals should loudly, drum on trunks of trees with their hands, shake branches and run about, creating absolute pandemonium. Whereas calls adapted for long-distance communication are generally stereotyped and therefore discrete, those employed in more intimate close-range situations tend to be graded and often accompanied by visual signals. Even the least sociable of mammals use vocalizations as a means of communication with conspecifics when mating. The male yellow-footed marsupial mouse (*Antechinus flavipes*), for example, makes a sound resembling '*cha-cha-cha*' during the precopulatory chase and male hamsters (Cricetus), fennec foxes (Fennecus zerda) and dwarf mongooses also have mating calls.

The male domesticated cat growls and the female screams when mating, and the male European polecat makes a clucking sound, while the female screams; female white-collared mangabeys (*Cercocebus torquatus*) also produce a mating call. Aggressive behaviour is very often preceded by vocalization. A threatening Herero musk shrew (*Crocidura flavescens herero*) will emit a single sharp metallic squeak. Common shrews (*Sorex araneus*) scream at their opponents, while baboons bark or roar; male mammals as diverse as deer (*Cervus elaphus*), lions and Northern elephant seals (*Mirounga angustirostris*) roar at male rivals when defending their harem of females. Threatening growls, familiar in domesticated dogs and wolves, also occur in Olympic marmots (*Marmota olympus*) and meerkats (*Suricata suricatta*).

The marsupial native cats (*Dasyuridae*) make a hissing or panting sound when threatening; hissing is also common in many carnivores such as polecats and among rodents. In some social species, individuals may attempt to steal food from one another; in the case of the meerkat, the thief is threatened with growling whilst, in common marmosets (*Callithrix jacchus jacchus*), the owner of the food produces a vocalization which has been described as *'erh-erh'*. The losers of aggressive encounters submit or glee but they may also vocalize, and the scream, squeak or squeal is a common sound associated with submission. This type of vocalization is characteristic of fear in mammals as diverse as rats (*R. norvegicus*), asses (*Equus*

Fig. 4.10. Sorex (Shrew).

hemionus), polecats, and many primates, including the great apes. Screaming is generally associated with gear in high-intensity agonistic interactions, usually where physical attack or fighting is involved.

On the other hand, whining, whimpering or squealing which are common juvenile vocalizations, in adults are generally associated with submission to a dominant who is in proximity, or threatening but not actively attacking. These vocalizations are common in primates and canids and they may also occur in a sexual context, female primates often squeaking or whimpering when approached by a male. The general message conveyed by such vocalizations is of a lack of aggressiveness and a willingness to submit to the other individual; an example is the yelp or whine of the coyote (Canis latrans) which indicates subordination. An important type of vocalization in many gregarious species is the contact call, which enables group members to keep in close proximity when visibility is poor. Such calls are easy to locate since they are usually low-pitched (for example the grunting of pigs) or frequency-modulated (like the contact trills of marmosets). Because the call draws attention to the exact location of the signaller, contact calls in prey species are usually of low amplitude. Female mammals often have a contact call which enables them to stay close to their young.

For example, mother and infant Northern elephant seals keep in contact by calling. When the mother comes ashore she calls to her pup which, after 2-3 days, learns to recognize her individually by means of this vocalization. Other herd-living animals, such as artiodactyls, have mother-young contact calls. As alien young are rejected by females, the infants soon learn to recognize their own mothers so that their capacity to identify their parent is both positively reinforced by suckling and negatively reinforced by attacks from other females. In some species of mammal the infant vocalizes when separated from its mother or accidentally deposited outside the nest, as in the case of the ultrasonic 'lost calls' of rat and mouse pups.

A young marmoset (*Callithrix jacchus*) separated from its family group produces a plaintive lost cry which attracts a carrier to pick it up, and the 'hoo-sigh' is an equivalent juvenile lost call produced by young white-handed gibbons (*Hylobates lar*). This brief description has covered only the common set types of call found in mammals but does not include all the call types or event the contexts in which calls are made in the social Canidae and higher primates. *Goodall*, for example, lists twenty-four different types of vocalization for the chimpanzee (*Pan troglodytes*) and it is clear that other apes are similarly well endowed. Even for the meerkat *Ewer* listed ten vocalizations : contract sound, settling-down sound when going to sleep, satisfaction noise when feeding, fear call, fear and aggression call, alarm bark, dissatisfaction call and three distinct types of threat, namely, growl, explosive spit and harsh repetitive scolding. As has been seen, one of the problems in using sound as a means of communication is that it may draw the attention of a predator to the caller.

Snowdon and *Hodun* showed that the most locatable of three contact calls in pygmy marmosets (*Cebuella pygmaea*), the j-call, is used in long distance communication. This call is high-pitched (9.2-13.2 kHz) but amplitude modulated, and Snowdon and Hodun took the view that, when jungle background noise is taken into account it occupies a unique acoustic channel above the highest frequency perceivable by predators. Even though high-pitched sound attenuates rapidly, the call is effective because of the restricted diameter (about 100 metres) of the pygmy marmoset's home range. *Snowdon* and *Cleveland* also showed that, in this species, individuals can be recognized from their contact calls.

Visual Communication

The more primitive mammals tend to be nocturnal or live in burrows, and elaborate visual signals tend therefore to be found mainly among advanced, diurnal species. Most mammals are rather dull in colour and, unlike birds, colouring typically plays little part in their communication system, although some diurnal primates may be highly coloured (e.g., mandrill). When *Darwin* (1873) drew attention to what he termed '*the principle of antithesis*', which he illustrated by reference to the aggressive and submissive postures of the dog, he noted that 'when a directly opposite state of mind is induced there is a strong and involuntary tendency to the performance of movements of a directly opposite nature'.

It is now clear that the principle of antithesis is only one aspect of the complex organization of the visual signalling systems of highly social mammals. Visual signalling may involve the use of any part of the body. Mammals may raise their hair, move their ears, alter their facial expression, change the angles of head, neck or tail; their limbs may be bent or extended and their genitalia may be erected or swollen. Sometimes these actions are further emphasized by markings on the pelage, as on the face of a tabby cat or tiger. In some cases, special structures such as horns or trunks may be used to signal.

Most visual signals are ephemeral, signalling the mood of the moment, but a few may convey a longer-lasting message, for example, the swollen female genital region of some old world primates and the antlers and mane of the rutting stag. Perhaps the best way to consider visual signals is to examine the way in which the different parts of the body may be used for this purpose.

Head

Even primitive mammals, which possess well developed teeth, may signal aggression or defence by opening the mouth and displaying the dentition. In some species of mammals such as wolves and baboons, the lips can be drawn back in snarl. The extremely mobile lips of the great apes (Pongidae) enable them to show a wide variety of facial expression; they may alter the shape of the mouth without showing the teeth, or may reveal upper teeth, lower teeth, or both, each expression being indicative of a different emotional state. Facial expressions change when vocalizing as in the case of the scream in polecats and primates such as the common marmoset. In some cases the tongue is used for signalling; cotton-topped tamarins (*Suguinus oedipus*) use the tongue in-and-out movement in situations inducing mild alarm.

In Old World monkeys, lip-smacking has probably been derived from grooming movements and also normally occurs in a friendly context. Pouting is another movement made with the lips in many

Old World monkeys and apes; it may have derived from suckling. Another expression, the open-mouth play face, occurs in carnivores and primates. The mouth is widely open, revealing the teeth, but there is no accompanying snarl or frown. The eye, which may be opened or closed, is another facial feature which can be altered, even in the most primitive mammals with rather immobile faces. In many diurnal primates such as monkeys, the eyes can be half closed or slitted. The closure of eyes often occurs in an aggressive context, as it does, for example, in fighting mice where the eye nearest to the opponent is usually closed, probably for protection. *Chance* suggested that closure of the eyes in an aggressive context also serves to cut off the fear-inducing stimuli emanating from the opponent and coined the phrase 'cut-off' to describe this phenomenon.

In addition to changing eye shape, mammals with forwardly directed eyes can lower the eyebrows to produce a frown. This is common in social carnivores and higher primates, and is indicative of anxiety or threat. In carnivores, horses and the more primitive primates the mobile ear pinnae are often used for communication. Submissive dogs and threatening horses lay back the ears. Common marmosets can flicks the ears when aggressive; the movement is emphasized by a tuft of white hairs surrounding the ear which

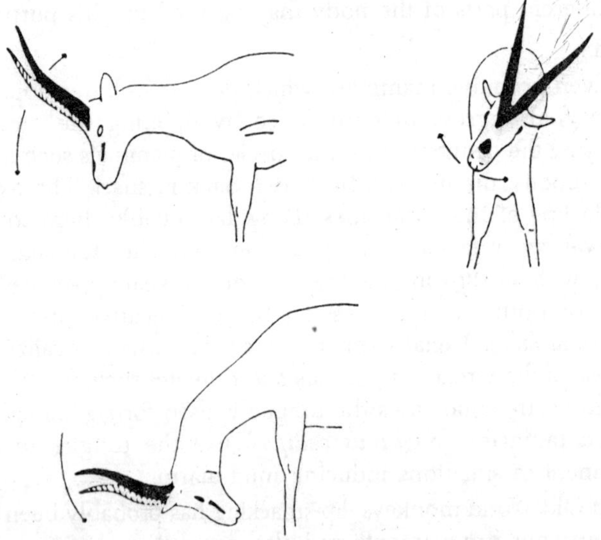

Fig. 4.11. Threat ritualized butting movements carried out by a male Thomson's gazelle.

contrasts with the generally back colour of the head. The cotton-topped tamarin raises or lowers the hairs of the crown of the head when threatened or alarmed. Sometimes the head bears special structures in the form of horns of antlers which emphasize the display, as for example Thomson's gazelle (*Gazella thomsoni*), which tosses its head to symbolize butting movements.

The head of the Thomson's gazelle is also patterned to emphasize the effect of the horns. Primates which have good colour vision may use colour or pattern to emphasize displays, for example, the blue eyelids of the crab-eating macaque (*Macaca fascicularis*) and the red and blue facial marking of the mandrill (Mandrillus sphinx). The red-bellied tamarin (*Saguinus labiatus labiatus*) has striking white lips which help to accentuate its open-mouth displays. The facial expressions of social canids and higher primates bear marked resemblances to one another, and van Hoof has compared 14 facial expressions for wolf, coyote and dog with those of a typical primate. There do seem to be considerable resemblances and it seems probable that these are a result of parallel evolution in the two lines from simpler flight, approach and defence intention movements.

Body

The mammal may use its body as a means of communication in three main ways, i.e. by piloerection, by posture or by movement of the whole or parts of the body. Often two of these methods of display are used in combination, and the display usually serves to emphasize particular parts of the body. The body fur in most mammals can be raised as a whole or in patches, one effect being to make the animal appear larger and so deter an opponent from attack. The familiar arched defensive posture of the domestic cat illustrates this point. Piloerection displays have undoubtedly evolved from the natural response of the animal to a stressful situation when production of adrenalin, which prepares the body for flight or fight, also results in the erection of the hair through the contraction of the arrector pili muscles. This reaction may, as in human, have little obvious functional value or it may serve as the basis of a display.

In some cases, erection of the hair may be emphasized by body patterning as, for example, in the case of the striped flanks of some ground squirrels. The aardwolf (*Proteles cristatus*) is of special interest because it is an anteating hyaena whose teeth are reduced to small pegs. In colouring it resembles, and possibly even mimics, its relative the spotted hyaena (*Crocuta crocuta*); its defensive threat display is also similar but, unlike *Crocuta,* which, of course, has a full and impressive dentition, *Proteles* does not reveal its teeth, the normal accompaniment to piloerection displays in carnivores. Bodily posture may serve as a form of social communication and *Grant* and *Mackintosh* described a whole series of postures in rodents, such as upright defensive posture and sideways posture, which appear in a number of different social contexts.

In many ungulates, visual bodily signals may be extremely conspicuous. For example, the male ibex (*Capra ibex*), like many other Caprinae, stands on its hind legs when threatening, while a subordinate Marco Polo sheep (*Ovis ammon*) kneels to the dominant. Juvenile or subordinate wolves roll over on to the back with inguinal region displayed and front paws retracted. The red-bellied tamarin sits up on its hind legs, displaying the golden-red belly. Many species display the genitalia and, in some monkeys and the chimpanzee, the anogenital area swells up and may also become brightly coloured during the female's periovulatory period (i.e. just before and just after ovulation). Female chimpanzees present their genital areas to males, and this may also serve as submissive behaviour in many primates, especially macaques and baboons. Presenting the rear to other individuals is practiced not only by females but also by males and is a submissive signal.

A dominant male may mount the subordinate and thrust. Display of genitalia, however, may signify an aggressive or dominant status. In common marmosets for example, the tail is raised and the genitalia displayed by presenting the hindquarters towards rival groups. In squirrel monkeys, the male displays his erect penis and the frequency of penis display is positively correlated with the male's rank. This, like female presenting in Old World monkeys, is an intragroup display. *Wickler* discusses the significance of genital displays in primates in relation to human behaviour. Some male primates have coloured genitalia, which emphasize the display, for example the vervet monkey which has a bright orange prepuce and blue scrotum.

In the case of the gelada baboon (*Theropithecus gelada*), the pattern on the genitalia is repeated on the animal's chest so that a similar signal appears even if the animal is in the sitting position, which it commonly adopts when feeding. Sexually receptive females commonly present the genitalia to the male, for example, rats which adopt a special position known as 'lordosis' in which the hindquarters are raised, the genitalia displayed and the tail deflected to one side. Female lions also display their genitals by raising the tail and walking alongside the male with the pelvic region raised. *Walther* has pointed out that ungulates may use the whole body to manipulate the behaviour of others. Grant's gazelle (*Gazella granti*) males herd their harem of females by using a series of movements. From this diagram it can be seen that the male can exercise considerable control over the direction of movement of females.

Tail

In many mammals, where this organ is ell developed, the tail may be used to communicate with conspecifics. Not only can its position be varied but it can, to varying degrees, be piloerected. In tamarins (*Saguinus*) the tail can also be coiled. Ring-tailed lemurs smear the tail with scent from the wrist glands and hold it aloft. As it is ringed with black and white bands, it is very conspicuous and so acts as both a visual and olfactory signal. Aggressive dogs and wolves also hold the tail aloft while mice thrash it from side to side ('*tail rattling*'). The extent to which these movements are of communicative values is not known. Clearly the division of visual signals into head, body and tail is highly artificial, as often more than one part of the body is simultaneously involved, particularly in highly social species issuing graded signals. Movement is very often used to reinforce the bodily signal. Threats may be emphasized by charging at the opponent, as in baboons, or circling it, like the African buffalo (*Syncerus caffer*).

Low-ranking animals may slink past dominants, or look away, giving a strong impression of nervousness, while high-ranking individuals appear confident and decisive in their movements which are unrestricted by the presence of others. Some species recognize individuals visually. This is quite clearly true of human beings who are capable of picking out a familiar face in a crowd of a thousand strangers. This facility is probably shared with other higher primates, at least with the more socially complex of the Old World monkeys and the great apes (Pongidae). Chimpanzees are particularly easy for us to recognize as individuals from their facial appearances, so it is likely that they also use the same cues. This ability may not be confined to primates, but hard data are not available for other mammalian orders.

Tactile Communications

Suckling is the earliest form of tactile communication to develop in a mammal. The sucking movements of the infant's lips not only induce milk let-down but also stimulate further milk production. Although the amount of information on other forms of tactile communication, with the exception of allogrooming, is restricted, it is clear that nocturnal, crepuscular and burrowing animals may rely heavily on this type of signalling. While, the function of biting in fighting is undoubtedly to inflict injury, it may also serve as a form of social communication. Snap biting is a common form of warming signal in carnivores and primates; inhibited biting also occurs in both social play and sexual behaviour in many species of mammal.

Male marsupial 'native cats' (Dasyuridae) such as the Tasmanian devil (*arcophilus harrisii*) and placental carnivores such as the black bear (*Ursus americanus*) grip the neck of the female preparatory to mountain, while the lesser hedgehog tenrec (*Echninops telfairi*) bites the spines on the female's back to stimulate her. The male greater hedgehog tenrec (*Setifer setosus*) actually bites the female during intromission, and she responds by trying to evade the bite, thus raising her back and improving the presentation of her genitalia to the male. Body rubbing, which may also involve the transfer of body odours, often occurs in a sexual or affectional context and is common in tenrecs. Many mammals have highly developed vibrissae which may be used in social communication. Nuzzling is certainly common in sexual and affectional behaviour so that the vibrissae, in species where they are well developed, are likely to play some role in the sensation created. Licking of the genitalia often occurs prior to mating, but is also a common response to a member of the opposite sex in dogs.

It has been suggested that genital licking has been derived from the genital grooming of infants, but it may also involve a chemical message. Dogs and wolves lick the muzzles of dominant individuals, behaviour which may have evolved from the infantile pattern of food soliciting. Kissing is a form of greeting in prairie dogs and great apes; for example, adult chimpanzees kiss when greeting and organs utan mothers kiss their infants. It is thought that kissing in primates may have been derived from oral allogrooming movements. Allogrooming, in addition to its cleaning function, is a very important form of tactile communication in

Fig. 4.12. Canis dingo (Dingo).

most social mammals. Rats, however, show a rough form of allogrooming termed '*aggressive groom*' which appears to be indicative of the aggressive mood of the groomer. Where a pair-bond develops between members of the opposite sex, allogrooming and other forms of tactile communication may be of great importance. The pair commonly sit in close physical contact (huddle) and the monogamous South American titi monkeys (*Callicebus moloch*) often sit with tails intertwined. Social play involves a great deal of tactile behaviour, especially wresting and rough-and-tumble play in primates and carnivores and the gentle butting of lambs and other horned artiodactyls. There is clearly considerable scope for further research on tactile communication, particularly in the more primitive mammalian orders and in relation to the role of the vibrissae.

Agonistic Behaviour

Agonistic behaviour is that which is associated with attack ad defence. Because an individual's reproductive potential is greater than the carrying capacity of a stable environment, intraspecific competition for the limited resources available is inevitable. Some of this competition takes the form of defending assets which can be monopolized, and this is the commonest situation in which agonistic behaviour is shown. To acquire and defend a resource, an individual must be capable both of obtaining it in he face of rivals and then of defending it in the event of a challenge. The assets which are most commonly defended are a mate, a food supply, living space, a resting site safe from predators, or the superior status which gives priority of access to such resources. In a situation of conflict an individual can either challenge its rival, or retreat. The advantage of challenging lies in the chance of winning the resource, which will be referred to as the payoff. The cost of challenging can be assessed in terms of the energy utilized in the conflict and the risk of serious injury. For the stronger opponent, challenging is the best strategy; for the weaker, retreat alone will provide the maximum benefit. It unequal contests it is clearly beneficial to have some method of assessing relative strengths which will allow the weaker individual to withdraw without physical injury.

Such an assessment will benefit both individuals if physical conflict can be avoided, and commonly occurs where a clear discrepancy in fighting potential exists between tow individuals. The stronger individual displays and the weaker withdraws. Even in situations in which two individuals are equally matched it would obviously be advantageous if a conflict could be resolved, without resort to have been violence, simply by display. Non-injurious contests of this type have been termed ritualized fighting and, until the early 1970s, many ethologists believed that ritualized fighting had evolved because it benefited both partners and hence ultimately the species. However *Maynard-Smith*, *Price* and *Parker* carried out a theoretical analysis of aggression using game theory which cast serious doubts on such an interpretation, and their views have been summarized by *Dawkins* and *Krebs* to which readers are referred for a more detailed account.

In essence their argument is as follows: if all individuals in a population settled disputes by ritualized fighting and an individual appeared who employed escalated injurious fighting (a 'hawk'), the 'hawk' would win every time. Thus, the strategy 'mouse' is not an evolutionarily stable strategy (or ESS) for it can always be defeated by a 'hawk'. (It will be recalled that an evolutionarily stable strategy is defined as a strategy which, if adopted by most members of a population, confers greater reproductive fitness than any alternative strategy). To continue, as a 'hawk' would always succeed in a conflict with a 'mouse', the genes for 'hawk' would spread throughout the population. The time would come, however, when the probability of a hawk meeting another hawk in a conflict would increase to such a point that serous injury among hawks would be common; mice would then come to be at an advantage. Thus 'hawk' is not an Ess either and in this example, assuming that he cost of injury was greater than the benefit of winning, the final ESS would be an equilibrium point with a particular proportion of mice and hawks.

The exact percentages of hawk and mouse can be calculated by game theory for various values of costs and pay off for the two strategies. *Maynard-Smith* appreciated that this scenario is simplified and that, for example, an individual might display conventionally against a mouse but escalate when confronted with a hawk. He termed this strategy 'retaliator' and showed that it can be an ESS in situations where the cost of injury exceeds the benefit of winning. This type of analysis is helpful to our understanding of aggression by providing realistic models which can easily be quantified. The examples so far considered have made the assumption that the two individuals are evenly matched in the sense that each rival has a 50:50 chance of winning.

In nature, however, conflicts are often asymmetrical; this may be because one individual is stronger (has greater resource-holding potential, or RHP), or one has greater need of the disputed resource, or because one individual is already holding the resource while its rival is a challenger attempting to take over. In such asymmetrical contests, information regarding the resource-holding potential of the combatants may be available. *Geist* found that, in mountain sheep (*Ovis dalli*), a strange ram's position in the dominance hierarchy is partially determined by relative size, especially horn-size, and he observed only flights between rams of approximately equal size. Thus, it must be assumed hat a ram is capable of assessing, on the basis of body and horn-size, the RHP of its rival in comparison to its own. Where individuals are unable to detect differences in RHP, they might be expected to behave as if they were evenly matched. This would lead either to an impasse or to damaging fighting.

Maynard-Smith and *Parker*, however, have shown that an arbitrary rule could be used to settle such disputes amicably. The rule, which they suggested, was that where two individuals appear to be equally matched the resource-holder wins, and it is apparent that such a convention would save time in display and avoid injury. What is surprising however, is that they found that the application of this arbitrary rule is an ESS having a higher payoff than either 'mouse' or 'hawk' strategies. The rule 'resource-holder wins' is common to most territorial mammals but may also occur in competition for females. In a lion pride, for example, a male who is already consorting with a female is seldom challenged and Kummer, in an experimental study of the formation of sexual bonds in hamadryas baboons found that a male will not challenge a pair which he has observed forming only fifteen minutes earlier. Prior ownership of a female confers the right to retain her, but it applies only where the resource-holding potential of the two males is similar. If the first male is obviously smaller than the second, the latter may challenge the owner and acquire the female.

Kummer also found that conflict developed when a male was removed from the troop and his females taken over by other males. On returning to the troop, the original male challenged the new owner and attempted to regain his females by force. Whether or not escalated fighting occurs is also dependent upon the scarcity of the resource, conflict being more likely if a resource is highly restricted, when the payoff will be high. Escalated fighting is also more likely to occur in species in which the risk of serious injury is low and these two factors must be taken into account in any investigation of fighting.

Weapons and Fighting Techniques

The majority of mammals possess weapons which may be used in attacking other individuals. The primitive mammal has a battery of at least 44 sharp teeth adapted for piercing the carapaces of insect prey or killing small vertebrates; these make effective weapons against conspecifics. Biting is the most primitive form of aggression and is employed by most mammals. Where biting is the major offensive strategy, the mammal must both succeed in biting its opponent and itself avoid being bitten. Mice (*Mus musculus*) attack parts of the opponent's anatomy remote from its jaws such as the back and flank, followed by very rapid withdrawal, thus employing a series of sharp nips.

In contrast, male Northern elephant seals make brief slashing bites at the opponent's head, neck and pendant nose; though deep wounds may be inflicted, these heal rapidly. Most terrestrial carnivores also use their teeth in fighting, in many cases biting the neck of the opponent as, for example, polecats. Cats also use their claws and primates may hit the opponent with the hand. Ungulates may use their hooves but some have, in addition, evolved special weapons such as the horns and antlers of ruminants. Geist regarded these appendages as having evolved for intraspecific aggression, as opposed to defence against predators, and *Leuthold* gives four reasons for this view:

1. The weapons are rarely used against predators.
2. For attacks on predators straight pointed horns would be the most effective most species have branched antlers or curved horns.
3. The horns or antlers of females are usually either absent or less well developed than thosc of males.
4. Cervidac (deer) usually only develop antlers for the rut.

When fighting, horned artiodactyls generally lock their antlers together and push one another, making the fight more equivalent to a wrestling match than a sword fight. Nonetheless, it is possible for one individual to make a broadside attack and inflict a severe injury, if it can get past its opponent's guard. In species of ungulate where both sexes are horned all the year bound, such as many antelopes, *Kiley-Worthington* takes the view that their horns prove very effective weapons against predators. She does not, however, dispute that they are of importance in intraspecific aggression and cites the example of the eland (*Taurotragus oryx*) in which the female has long thin horns suitable for delivering quick stabs, while the male has more robust weapons suitable for horn-wrestling with other males. Elephants fight by pushing one another, using the tusks to prevent the opponent's head slipping sideways; rhinoceroses club their adversary's head with their nasal horns.

Zebras (*Equus burchelli*) bite and also kick with their hooves, rising up on their hind legs. Amongst the most bizarre forms of fighting is that of the giraffe (*Giraffa camelopardalis*), which is termed *necking*; the animal swings its whole neck and head towards the opponent, aiming for the shoulder. One of the most remarkable weapons possessed by any mammal is the male narwhal's (*Monodon monoceros*) single long tusk which is also believed to be used in intraspecific fighting. Many mammals which fight have thickened skin in regions where they are likely to be attacked, e.g., rhinoceroses, elephant seals and polecats. Fighting, apart from that of pinnipeds, is difficult to observe in the field because it rarely occurs, and many agonistic encounters which are observed may be between individuals who have met previously and already established their status in relation to one another.

In many species fighting is of short duration and thus difficult to analyse, but escalated fighting in male polecats, which may last for as long as nine minutes, has been subjected to a detailed cine analysis in the laboratory. The fighting technique in this species consists essentially of biting the opponent and holding on for as long as possible. Biting occupies approximately 40% of the time, and the individual which can sustain its bites longest wins.

Table 4.2. Orientation and duration of bites in polecat's fighting for 917 bites with a total duration of 41.96 seconds (60418 frames of cine film at 25 fps).

	Bites	
	No.	*Mean duration (seconds)*
Face	14	2.0
Neck	832	3.0
Chest	1	2.7
Shoulder	50	1.8
Back	49	0.8
Flank	31	1.5
Belly	1	0.8
Anal region	22	1.0

Over 80% of the polecat's bites are directed to the neck of its opponent, a region well protected by skin and subcutaneous fat. It might appear that polecats bite their opponent in an area where they can do least damage, in line with Lorenz's suggestion that natural selection favours non-injurious fighting which promotes species survival. Further examination of the data, however, shows that bites to the neck are retained far longer than those on other parts of the body. (See Table 4.2) As winning is associated with the duration of biting it is clearly to the individual's advantage to bite the neck in preference to other regions. The fact that the neck is well protected is therefore hardly surprising. These data also support *Maynard Smith* and *Parker's* view that escalated fighting is only likely to occur where minimal injury is likely to be inflicted.

The Role of Aggressive Displays

One of the few species of mammal for which a detailed analysis of aggression is available under field conditions is the red deer which has been extensively studied by *Clutton-Brock* et al. In the rut, each stag attempts to acquire access to a group of hinds (a harem) which he then defends against rivals. In situations in which a stag is likely to be challenged he shows threatening behaviour in the form of roaring and walking parallel to his opponent. Should the other male continue to challenge, however, the antlers are used and a fight develops. Fighting stages lower the head and ram the opponent with the antlers.

As the rival faces his attacker, their antlers lock together and a pushing match ensues. Each stag attempts to get uphill from his opponent and to push him backwards. The loser finally withdraws and takes to flight. Pursuit is rare because attacking involves lowering the head which reduced the male's speed. If, however, the loser slips or falls, the winner will take the opportunity to jab the flank of its fallen rival, often wounding it severely. Fighting success is positively correlated with reproductive success and, in turn, with antler size. Success in fighting can therefore lead to a considerable gain in a male's fitness, but it also involves risks, as there is a 1:17 chance of its being seriously wounded. As a seriously-wounded stag has little chance of survival, it is apparent that the optimum strategy is to avoid contests with superior rivals or those in which there is a high risk of injury. *Clutton-Brock* et al., investigated the displays which occurred prior to fighting to determine their significance and likely function. They found that stags roar most in conditions in which they are likely to be challenged and that the pitch of roaring is related to the size of the animal, larger stags having lower-pitched roars. Roaring provokes other stags to reply, and the frequency of the replies is positively correlated with the frequency of the roars heard.

Generally stags roar alternately, each facing its opponent. If one contestant withdraws at this stage, the winner is the individual, which has the maximum number of roars per bout. Human observers can predict the outcome of a contest by comparing the two stags' roaring rates. Clutton-Brock et al. Concluded that roaring appears to function for the two stages as a means of assessment. If the protagonist remains after the roaring contest, the contestants generally commence parallel walking. This occurs most commonly when opponents are well matched. The fact that there is a strong correlation between the duration of parallel walking and that of subsequent fighting again indicates a trial of strength; very long parallel walks, however, are less likely to be followed by a fight. These data support the view that roaring and parallel walks provide a means whereby rival stage can assess on another's fighting potential. Stags seldom challenge older, larger individuals, so that roaring contests rarely occur if there is an obvious visible discrepancy in size; very powerful stages of approximately equal strength, however, generally do not fight after long parallel walks, thus avoiding serious injury.

Clutton-Brock et al. discuss how roaring and parallel walks might be related to fighting ability; they regard parallel walking as a test of stamina and they put forward the view that roaring and fighting both involve the same thoracic musculature. These results can be interpreted using *Maynard-Smith* and *Parker's* analytical approach. Contests are usually asymmetrical; one individual holds a harem while its challenger does not. The rivals confront one another and assess each other's fighting ability, without actual combat, either by size discrepancy or vigour of display. This method of assessment is probably relatively immune to cheating. If the two individuals come to regard one another as equals they then adopt the convention 'resource holder wins'. The above discussion has considered only cases where individuals simply advertise their strength; in other situations individuals may avoid showing their weaknesses. Young male stags roar less than older ones when in possession of a harem, thus avoiding the attention of older, larger stags.

Packer has shown that, in male olive baboons, fighting ability is related to canine wear, and open-mouth yawns are used to display the canines to opponents. Males with long, sharp canines yawn more frequently than those with worn or damaged ones. There have been few other quantitative investigations of the effects of different displays on the opponent, or the degree to which future behaviour can be predicted. *Chalmers*, however, provided some information on three displays in sooty mangabeys (*Cerococebus albigena*). It is apparent that yawning and staring convey an aggressive intent, while lipsmacking appears to be friendly. *Chalmers* also found that the orientation of the yawn influenced the behaviour of its opponent. Yawning directed towards an individual more often resulted in its flight than did yawning away from it. From the above discussion it is apparent it is apparent that, in some species, display may function to settle a dispute without physical conflict which could prove injurious. By advertising its fighting potential, the individual conveys information to its rival and at the same time may receive information about its opponent's strength.

Table 4.3. Subsequent behaviour of signaller and recipient after three facial displays. Data from a field study of mangabeys (Cercocebus albigena)

Behaviour		*Next behaviour*		
	Individual	*Attack*	*Remain*	*Flee*
Stare	Signaller	6	12	0
	Recipient	0	2	12
Yawn	Signaller	9	16	19
	Recipient	1	7	6
Lipsmacking	Signaller	0	21	9
	Recipient	0	14	0

Submissive and Defensive Displays

In situations I which there is a difference in resource holding potential (RHP) between two individuals, the weaker may display to indicate his non-combatant status. The two main type of display concerned are termed *submissive* or *defensive.* Submission indicates that the individual will not retaliate, even if attacked, and a common form of this behaviour is rolling over on to the back or crouching. Brown rats (R. norvegicus) also utter an ultrasonic cry which inhibits the dominant from attacking.

Submissive canids lie on the back and expose the inguinal region, sometimes also showing a submissive grin. Submissive displays are commonly shown by the young towards older group members, by females towards males and by very low-ranking adults towards high-ranking dominants. Defensive displays, often referred to as 'defensive threat' signify that the animal will not spontaneously attack its opponent but may retaliate if attacked. It is commonly used where two closely-ranking individuals are in proximity, or where a lower-ranking individual is holding a disputed resource. In many species defensive threat involves displaying the teeth as, for example, in some marsupials, primates and carnivores.

Defensive threat and submission occur in situations where the signaller wishes to avoid conflict or the resumption of a previous fight, the difference in RHP between the interactors being greater in the case of submission. In species which live in social groups, memory of the outcome of previous conflicts with a particular individual must be the important factor in determining the nature of the display and it is in the more sociable mammals that the most elaborate forms of agonistic displays have evolved.

Agonistic Displays and Their Underlying Motivation

Aggressive, defensive and submissive displays have been defined and discussed independently. In highly sociable species, however, forms of display intermediate in form between these categories have been identified. They have generally been interpreted in terms of a conflict between the motivation of aggression (urge to attack) and fear (urge to flee), but few quantitative data are available to support this view as regards mammals. *Maynard-Smith*, *Dawkins* and *Krebs* and *Caryl* have, as already stated, questioned its validity by pointing out that, in an agonistic contest, an opponent would be at a disadvantage if the provided accurate information about his next move, and that game theory would appear to favour a bluffing strategy giving the impression of greater confidence than might be the case.

Hinde argued that a display may result from conflicting motivations but convey only very generalized messages such as 'I shall attack or stay' of 'If you do x I am more likely to attack than escape'. In the absence of relevant quantitative data for mammals, it is not currently possible to resolve the conflicting views regarding the motivation of these displays, the ways in which they are interpreted by a rival, or their reliability as predictors of subsequent behaviour. *Hutson* carried out a quantitative study of threat in the Australian carnivorous marsupial, the kowari (*Dasyuroides byrnei*). He took into account both the sequences of behaviour shown by the signaller and the response of the opponent. Two types of threat display were shown by this species towards conspecifics. One was the pant-hiss which was associated with a high probability of attack and significantly elicited flight from the opponent.

The other was shown by a defensive individual and was associated with flight or retaliation, but not with spontaneous attack. In view of the considerable variation in intensity of the calls. *Hutson* took the view that the data were compatible with the conflict hypothesis of threat display, but did not provide data to show what effect intermediate displays might have on the opponent. He found that distance from the opponent was important, as was the

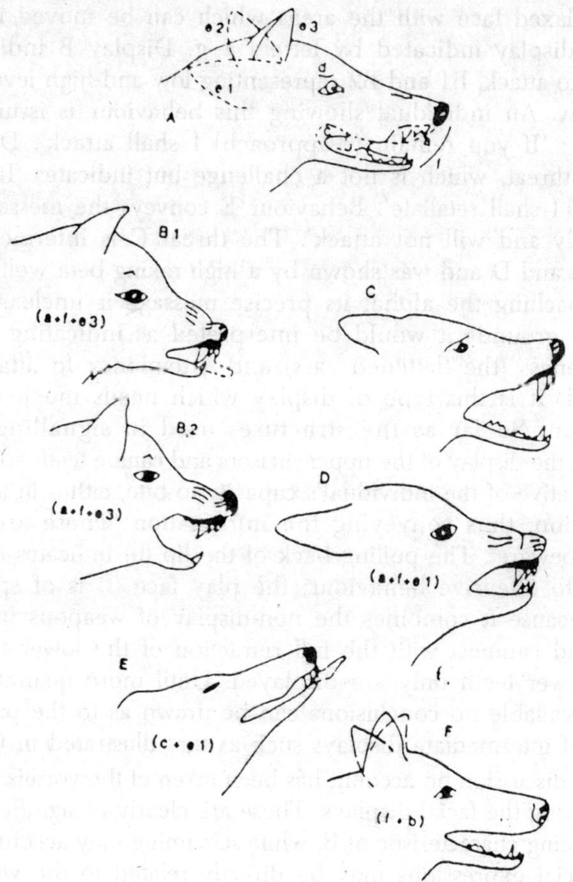

Fig. 4.13. Some facial expressions of the wolf. A–parts of the face utilized in expressive movements, (a) upper lip, (b), (c) lower lip positions, (d) eyebrow (frown), (e) ears, (f) opening of the mouth; B–differing intensities of confident threat; C–defensive threat, D–threat by beta male to dominant; E–friendly submission; F–open-mouth play face.

context in which the display occurred–for example, whether the animal was approaching or fleeing from its rival. A species whose aggressive displays have generally been taken to indicate conflicting tendencies to attack or flee is the wolf, together with other related canids such as the dog, coyote, fox and jackal.

We shall briefly consider some of these displays in terms of the information which they are believed, by a number of experienced observers to convey to a rival. Figure shows some facial expressions of the wolf taken from photographs published by *Fox. A* shows the normal relaxed face with the areas which can be moved in an agonistic display indicated by letters a–g. Display B indicates readiness to attack, B1 and B2 representing low and high levels of this display. An individual showing this behaviour is issuing a challenge : 'If you remain (r approach) I shall attack'. D is a defensive threat, which is not a challenge but indicates 'If you attack me, I shall retaliate'. Behaviour E conveys the message 'I am friendly and will not attack'. The threat C is intermediate between B and D and was shown by a high-raking beta wolf who was approaching the alpha; its precise message is unclear. On traditional grounds it would be interpreted as indicating both submissiveness (the flattened ears) and a tendency to attack if challenged. It is this type of display which needs much more clarification. So far as the structures used in signalling are concerned, the display of the upper incisors and canine teeth appears to be indicative of the individual's capacity to bite, either in attack or retaliation, thus conveying the information 'These are my weapons, beware'. The pulling back of the lip (b) indicates a low tendency to offensive behaviour; the play face (f) is of special interest because it combines the non-display of weapons (upper incisors and canines) with the full retraction of the lower lip so that the lower teeth only are displayed. Until more quantitative data are available no conclusions can be drawn as to the precise meaning of intermediate displays such as that illustrated in C.

In this discussion no account has been taken of the vocalizations accompanying the facial displays. These are clearly of significance, growling being characteristic of B, while screaming may accompany D. The facial expressions may be directly related to the way in which accompanying sounds can be produced. From the data currently available it is not possible to decide how precise a prediction of future behaviour an intermediate display may be. The view that it reflects underlying motivation however seems reasonable on face value. What needs to be sown is how these displays may benefit the signaller, which otherwise appears to be giving away important information about the future moves to the opponent. An aspect of mammalian agonistic displays which is often ignored is the fact that most graded displays in species living in social groups are directed towards familiar individuals. This being the case, the contestants already know one another and are aware of any previous asymmetries in their resource holding potential. Any display which reflects this asymmetry gives nothing away and may simply emphasize the persistence of a stable situation, thereby avoiding further challenges for the dominant. If the signaller were to bluff in this situation, it might well end up with an unwanted challenge.

An important social aspect is whether a resource is involved. If the resource is valuable to the signaller but less so to its rival, the weaker individual by adopting a form of defensive threat is effectively saying that it will retaliate only if the dominant individual attempts to take over the resource. In this situation the dominant may withdraw because the cost of acquiring the resource may be greater than its value.

Dominance Hierarchy

The concept of dominance hierarchy relates to a situation in which individuals can be ranked so that a higher-ranking individual may attack or show aggressive threat, while the subordinate avoids conflict by moving away or showing submission or defensive threat. In a social group the situation may exist where individual. A is dominant over B, which is in turn dominant over C and so on; this is a linear hierarchy but sometimes loops may exist so that A is dominant over B who is dominant over C who dominates A. Considering only dyadic relationships, therefore, the dominant individual can threaten or displace the subordinate with impunity. A dominance hierarchy is commonly established in one of two ways, either through fighting, in which the dominant wins, or through differences in age (age-graded rank). *Packer's* work on olive baboons provides an excellent example of both arrangements in a multi-male society.

Individuals born into the group have a status based on their age, older individuals being dominant over younger ones. Immigrant males, however, have rank which is not age-dependent but results from conflicts with other males. Japanese macaque females are unusual in that the status of sisters in inversely related to their age. In most social species males are dominant over females and *Dittus* found that, for each age group, male toque macaques dominated females. Some authors have questioned the validity of the concept of a dominance hierarchy on a number of grounds, for example, the fact that all measures of dominance (reproductive success, access to mates etc.) may not be correlated; the possibility exists that dominance hierarchies are a product of confinement and are of much less relevance in the field, and that subordination is the critical aspect of a dominance hierarchy.

If the concept is confined to situations where there is a consistent asymmetry in agonistic relations, then the term would seem to be valid because dominance hierarchies have been observed in the field in many primates, ungulates and wallabies. Mech (1970) examined the value of he concept of a dominance hierarchy in a wolf pack (*Canis lupus*). He correlated a series of parameters with dominance and obtained high correlations which gave reliable ranking, and concluded that a wolf pack can be regarded as a ranked society. Dominance hierarchies have been identified in the field in marsupials, rodents, lagomorphs, carnivores, artiodactyls, perissodactyls, elephants and primates. The degree of stability of dominance hierarchies may vary; in some species it only becomes apparent in the rut, e.g. Northern elephant seals, while in others it is a year-round phenomenon, although changes may occur during the breeding season. *Jolly* found that the male hierarchy of ring-tailed lemurs outside the breeding season differs from the which appears when females come into oestrus.

In wild yellow baboons, *Hausfater* found that the female dominance hierarchy is much more stable than that of males. Changes resulting from demographic factors such as death, emigration and immigration occurred every 13 days for the male, every 57 for females, while those due to agonistic interactions took place on average every 21 days. Two important questions which require further consideration are what benefits are associated with dominance and, if dominants do have an advantage, why subordinates remain in a social group. The general view among ethologists is that dominance confers greater reproductive success and *Dewsbury* has examined that data which purport to support this hypothesis. The usual means of measuring reproductive success are numbers of copulations by dominant, as opposed to subordinate group members, and number and duration of consortships between males and females in relation to rank.

One factor which should be taken into account is the stage of oestrus of the female when consortship takes place. High-ranking male olive baboons, for example, consort with a female only at peak oestrus, so that subordinate males are less likely to father offspring, irrespective of the duration of their consortship in relation to that of the dominant. *Drickamer* questioned the value of taking absolute measures of mating frequency in the wild; his data showed that dominant male rhesus macaques mated more frequently than subordinates but that the latter were less frequently seen and, when the data were corrected for time observed, the apparent difference disappeared. Order of mating may also be a significant factor in some species. *Levine* found that a female laboratory mouse which had received an equal number of ejaculations from two males bore more offspring from the first male; by contrast, female golden hamsters and prairie voles produced more offspring from the male who mated last.

Other factors which may influence mating success are sperm exhaustion in a dominant male which mates very frequently, as compared with a subordinate who rarely does so. *Dewsbury* concluded that there is good evidence for a correlation between dominance and reproductive success in carnivores, ungulates and rabbits, while the situation appears to be much more variable from species to species in rodents and primates. For females there is often a correlation between rank and reproductive success, measured by number of offspring produced. This has been demonstrated in captivity in deermice (*Peromyscus maniculatus*), house mice and red deer, and, in the field, in rabbits, in Japanese macaques by *Takahata*, and gelada baboons by *Dunbar* and *Dunbar*. *Dittus* found that subordinate toque macaques had less access to food, fewer refuges from predators and fewer mates and sleeping sites. This raises the question of how subordinate behaviour has evolved and what advantages it might confer on the individual.

By submitting to a stronger individual, the subordinate avoids an unequal contest and the danger of injury. It also saves the energy involved in a fight or flight which it can more profitably use to seek food elsewhere. If the dominant is also a relative, the subordinate gains some inclusive fitness by not committing its relative to a time-and energy-wasting conflict. Behaving in a subordinate manner allows an individual to maintain close proximity to high-ranking males of the group and to remain a member of the social group, thus gaining the benefits of group vigilance, group foraging and other advantages of social life. If a subordinate individual is younger than a dominant it may itself, in time, become dominant if it remains in the social group, or alternatively it may accept subordinate status in its natal group but, when another group is contacted, the individual may change groups and acquire a high status in the new group.

Vehrencamp has put forward a quantitative model comparing egalitarian societies with those showing despotic dominance (i.e., a single dominant which manipulates resources to its own advantage). She concludes that the extent to which benefits may be restricted to a few (dominant) members of a society is dependent upon the cost to subordinates leaving the group balanced against the benefits derived from group living and the degree to which they are genetically related. Agonistic behaviour is known to be an important factor in the regulation of numbers in r-selected rodents such as deer mice, and lemmings (*Dicrostonyx, Lemmus*), but recently Dittus has shown that it is also of paramount importance in a K-selected species, the toque macaque in Sri Lanka. *Dittus* found, in a three-and-a-half-year study, during which there was no significant population growth, that agonistic behaviour was the major factor in population regulation. Females usually remain in their natal groups, while males emigrate to another one on reaching maturity.

Table 4.4. Percentage Mortality Per Annum for Different Age, Sex, Classes for the Toque Macaque (Macaca Sinica).

		Percentage mortality per annum	
Stage	*Years*	*M*	*F*
Infant	0-1	39.50	52.60
Juvenile (1)	1-2	7.9	15.8
Juvenile (2), (3)	2-5	7.4	5.4
Subadult male	5-7	9.95	–
Adult male	7-30	0.45	–
Adult female	5-30	–	0.61

The sex ratio at birth is equal but that of individuals over five years old is biased towards females (1.56 F : 1 M). Of all individuals born, 85 of females and 90% of males died before adulthood. *Dittus* concluded that predation was negligible and that success in aggressive competition for food was the major factor which determined survival. Subordinates were supplanted from food in 36% of all threat interactions and survival was closely related to status. Males of any age group were dominant to females, so that the most vulnerable groups were young juvenile and infant females. There was, however, a high mortality in sub-adult males which were driven by the adult males to the periphery of their natal group and subsequently suffered the hazards of emigration of their natal group and subsequently suffered the hazards of emigration to another group. *Dittus*' data show that intraspecific aggression effectively results, under stable population conditions, in a pre-breeding mortality rate of between 85%-90% of all individuals born. From these examples it is apparent that status can have an important effect on reproductive success and, a mother's rank can even affect the sex ratio of her offspring.

Infanticide

There are, inmost social mammals, strong inhibitions against the killing of infants which are clearly the most vulnerable members of society. In some species, however, infanticide has been commonly observed, the best-documented cases being those of langurs and lions. Infanticide generally occurs after a harem has been taken over by a new male. The male increases his fitness by eliminating young which he has not fathered, thus reducing competition for his own offspring and bringing their mothers into oestrus earlier than would otherwise have been the case. In both common langurs and lions males only hold a harem for a brief period of up to two to three years, so that there must be considerable selection pressure favouring males which father infants very soon after taking over the harem.

The theoretical advantages of infanticide in langurs have been discussed by *Chapman* and *Hausfater* who pointed out the importance of the relationship between mean length of tenure of a harem and the evolution of infanticide. They compared a theoretical model with the values obtained from field studies and found good agreement. Infanticide is by no means universal in species with harems and, in fact, appears to be limited to a few species with a relatively short harem tenure. Females might be expected to oppose it concertedly, or to oppose group takeover by an infanticidal male. To take the female langur as an example, the loss of her young reduces her overall reproductive effort considerably, for she produces only seven or eight offspring in a lifetime and, through infanticide, loses a 6-15 month period of reproductive output if her 1-12-month-old infant is killed. In both lions and langurs, however, males are considerably larger than females so that individual female opposition is unlikely to be effective.

Packer and *Pusey* however, saw a group of three lactating female lions successfully defending their cubs against a group of alien males, although all three females were wounded. Lionesses also show an indirect strategy for reducing infanticide immediately after the pride males have been driven out rivals; females show increased sexual activity but low fertility for the next few months. This results in competition for the pride by male coalitions and takeover by the largest coalition. This, in turn, ensures a long period of tenure and hence a reduction in the frequency of post-takeover infanticide. *Hrdy* argued that females may favour aggressive males in a harem takeover, and that a more aggressive male is more likely to succeed in infanticide but is also likely to be a more effective protector than a more amicable one.

Until more is known of the relation between infanticide and aggressiveness in males, this view can only be regarded as speculation. A further factor which may be relevant is the proportion of females in a harem who are in oestrus and make takeover attractive to a male. Females may actively solicit an outside male, especially if they are daughters of the current harem-owner who are actively attempting to avoid mating with their father. Thus, it may be that the takeover is influenced by a majority of females who are not vulnerable to infanticide and therefore have less to lose and probably much to gain from takeover by anther male. For these females an infanticidal male may increase their fitness in two days, firstly, by removing infants who would complete with their own offspring and secondly, indirectly, through their production of infanticidal sons.

Both Sexes Solitary

In these species individuals of both sexes are spaced out throughout their environment and mutually avoid one another. Mating is opportunistic, a female advertises her oestrous condition and any resident males in the vicinity are attracted and may compete to mate with her. Spacing may be achieved by the use of chemical markers or, in some cases, by vocal display. Some solitary mammals are territorial, i.e., they defend their home range against intruders, and in many instances successful breeding can occur only when an individual holds a territory. Individuals which fail to acquire a territory, or have not yet done so, are usually nomadic. The shorttailed shrew is a well-documented example of a solitary species in which individuals defend territories against intruders of either sex. Females are only tolerant of males when sexually receptive. This social arrangement can be termed 1 *(a), intersexual or exclusive* territoriality. A situation which is also well known is that of 1 (b), *intrasexual* territoriality, where individuals of each sex defend their home range against intruders of the same but not the opposite sex.

Usually males hold larger territories than females, so that a male's territory may include those of several females. This arrangement will increase the male's opportunity of forming sexual relationships with any females within his territory, because the territorial male will detect their oestrous condition and, by excluding other males, decrease the probability of their mating with other males. Females may also show a preference for a territorial, as opposed to a vagrant, male as there may also be some element of female selection. Intrasexual territoriality is common in mustelids such as weasels, stoats, mink and others. The effect of male/male competition for territories which include more than one female is to produce a surplus of nomadic males which avoid male territories, unless challenging the owner, and search for vacant territories which often an opportunity to breed. Many solitary mammals do not defend their home range. They are, however, spaced out and may show a degree of hostility on meeting same sex conspecifics.

In *solitary non-territorial* species *(1c)* males usually have larger home ranges than females, particularly in the breeding season. Females attract males by indicating that they are in oestrus, and several males may congregate and compete for a female. This arrangement is characteristic of many small rodents, for example the field mouse *Apodemus*, and members of the cat family. It is probably the commonest social system in primitive mammals, which do not have readily defendable resources within their home range, but, surprisingly, is also shown by one of the great apes, the orang utan.

Monogamy

Monogamy occurs where one males and one female form an exclusive sexual relationship, but it is important, to distinguish at the outset between monogamy as a sexual strategy and monogamy as a social arrangement. *Wickler* and *Seibt* (1983) have pointed out that these two aspects are often confused, so that the concept of monogamy may be an ambiguous one. In discussing monogamy in mammals the social relationship will be regarded as more important than the number of mates which an individual may have. Fox example, in populations of mammals which are typically polygynous, there will be males who have managed to monopolize only one female. Such males have adopted a monogamous mating strategy, but only because they lack the competitive ability of opportunity to acquire exclusive mating rights to more than one female.

A monogamous social organization is typically one in which a breeding unit consists of one male and one female who share a unique social relationship from which third parties are excluded. *Kleiman* divided monogamy into two forms: (a) facultative, where individuals are so spaced out that only a single member of the opposite sex is available for mating, and (b) obligate monogamy, where a female cannot rear her young successfully, without direct assistance from conspecifics. *Fox* divided monogamous canids into three types based on their group structure. Both classifications have some merits so that a modified scheme based on both of them will be employed here (a) *Grade* I monogamous mammals are those in which a male and female defend a common territory but offspring to not remain with their parent(s) after weaning. Adults may be paired permanently or only during a restricted breeding season. The amount of contact between them also varies from one species to another. Elephant shrews, for example, spend very little time in contact, although they share the same territory, defending it against same sex conspecifics.

Common tree shrews have a similar social organization to elephant shrews, but a minority of males have territories large enough to include those of more than one female, so that they are sometimes polygynous. Another good example of a Grade I monogamous species, in which male and female keep in close contact, is the klipspringer (*Oreotragus oreotragus*) studied by *Dunbar* and Dunbar (1980). Both sexes defend the territory but, during the female's pregnancy and lactation, she spends more time feeding than the male, who generally spends more time being vigilant and thus allows his mate to feed in security. This division of labour enables the female to acquire the extra resources needed by the developing offspring. (b) *Grade II* species are permanently paired but delay driving out their young, so that adults are accompanied by more than one generation of offspring. Many species show direct paternal care, and juveniles and subadults may assist in the caring for offspring. Colonial Grade II monogamy is found in the beaver *(Castor fiber)*, where offspring remain in the family group until two years old, assisting the parents by cleaning the nest, collecting food for storage and caring for infants.

Grade II monogamous mammals, which form mobile groups, include some members of the dog family, the South American titi monkeys (*Callicebus moloch*) and the lesser apes (Hylobatidae). (c) *Grand III* species have multimale/multifemale groups where only a single male and female breed. Breeding is determined by an individual's position in a dominance hierarchy so that only the alpha male and female breed. Grade III monogamy can be regarded as status-determined monogamy; the social group is a multimale/multifemale one in which only the alpha pair breed. While non-breeding members of the group assist in the care of infants and co-operate in group hunting. Status-determined monogamy, which corresponds to the social organization of *Fox's* type III canids, is found in wolves (*Canis lupus*) and hunting dogs (*Lycaon pictus*). In view of the fact that canids are generally monogamous it seems most likely that this form of organization is derived from Grade II monogamy.

The pack, however, in *Lycaon*, is not simply a family, the alpha male is not necessarily the oldest male in the group, and dispersal is by emigration of the females out of their natal pack. Dominance relationships within the pack, unlike those of typical monogamous species, change from time to time. The wolf pack more closely resembles a family group, although there are both male and female dominance hierarchies. In view of the close social ties between members of a wolf pack and their hostility to unfamiliar conspecifics, they appear to form inbreeding groups with negligible gene flow between packs. Common marmosets also appear to show status-determined monogamy, as they live in multimale/multifemale groups with a dominant pair where only the alpha female breeds.

Polygyny

A polygynous social organization is one in which a male has exclusive sexual access to several females. This arrangement depends upon females being sufficiently tolerant of one another to enable them to live in close enough proximity to allow a single male to monopolize them. If female/female hostility is high they become so spaced out that the male is forced to adopt a monogamous or promiscuous sexual strategy. There are basically two types of Polygyny. The first occurs where a male's attractiveness to individual females enables him to assembles, or take over from anther male, a group of females, each of whom is socially bonded to the male but not to other females of the harem. The second type of polygyny occurs where the females in the harem are strongly bonded to each another and the female group exists as a persistent social structure in its own right.

Females are usually closely related and from time to time the group is monopolized by different males, each of whom supplants his predecessor. The second type of polygyny is associated with male dispersal from the natal group. The majority of gregarious mammals are polygynous. This class contains a large number of different forms of social organization which can be divided into four main types, harems, territorial males, spaced males and multimale/multifemale groups. *Harems* occur where one male, sometimes accompanied by an associated satellite male, has exclusive mating rights to a group of females. Four main types of harem have been described.

A *dispersed harem* occurs where a male holds a territory which encloses the home ranges of several females with whom he has exclusive mating rights. The females aggregate at a sleeping site or burrow but forage independently. In the case of the nocturnal bushbabies (*e.g. Galago demidovii*) females sleep together during the day in matriarchal groups, while non-territorial males are peripheral or subordinate in the alpha male's territory. Male bush-babies emigrate to breed, while strong mother-daughter affectional bonds exist which persist into adult life. Many colonial burrowing mammals show dispersed harems where females occupy burrows in a restricted area and a male includes the home ranges of a number of females within his territory; females forage independently, and mother-daughter home ranges show complete overlap *(e.g. Procavia johnsoni and Marmota olympus*). Males emigrate to breed and non-breeding males are peripheral. Subadult young are tolerated in yellow-bellied marmots (*Marmota flaviventris*).

Rookeries, or breeding assemblies, are found in highly mobile mammals which return to a restricted are to breed. Males compete to acquire a harem of female, although they also mate opportunistically with any female are which stays into their harem. There is no social cohesion, females are attracted to an area where they give birth and nurse their pups, while males defend a group of females which have a post-partum oestrus. This arrangement occurs in Pinnipedia, such as Northern elephant seals and Steller sea lions and in some bats (*e.g. Pteronotus parnelli, Mormoops megalophylla*).

Temporary mobile harems, where a male associates with a group of females, are typical of many cursorial mammals such as wild goats (*Capra hircus*) mountain sheep (*Ovis canadensis*) and red deer.

Permanent harems are found in other species. Fox example the male vicuna is territorial and defends a group of 4-18 adult females, while other males live in bachelor herds which are non-territorial. Both six-month-old sons and yearling daughters are driven out of the harem, with the result that its size remains relatively constant permanent mobile harems take one of two forms; either the male collects a group of unrelated females (non female-bonded harems) or he takes over a group of related females (female-bonded harems). The cohesion of the group depends upon the existence of a non-sexual male-female bond. This system occurs in some bats, such as *Saccopteryx bilineata* and *Phyllostomus* discolor, where the female composition of harems varies from time to time. The best known instances of non-female bonded harems are in primates, namely the hamadryas baboon and the gorilla.

In both species females leave their natal breeding unit and are attracted to a male who may or may not already have a harem. Bachelor groups do not occur in either of these species. In the hamadryas, the breeding unit is part of a larger band which includes some non-breeding adult males while male gorillas tend to remain in their natal group or become solitary and attempt to attract females from another harem. Female-bonded harems are the type found most frequently in mammals. The permanent group consists of a number of related females with strong affectional bonds between them. A male simply takes over the sexual access to the group from a previous male, defending the harem against other males. In most cases, males outside the breeding units live in bachelor groups, usually with a rank order. This system has been studied in common langurs by *Hrdy* and gelada baboons by *Dunbar* and *Dunbar*. It is also known to occur in mountains zebras (*Equus zebra*). Breeding units are part of a larger troop or herd on gelada baboons and feral stallions (*Equus caballus*), the latter being ranked independently of the size of the harem which they hold.

Territorial Males, Females More Sociable

In some mammals breeding males are territorial, but females are more sociable and have home ranges which overlap one or more male territories. A system termed by *Leuthold* 'territory and absolute hierarchy combined' was found in the white rhinoceros by *Owen-Smith.* In this species non-breeding males lived as subordinates in a dominant male's territory. Female home ranges overlapped those of males and groups of one or more adult females, with their juvenile and infant offspring, moved around together. In some species, non-territorial males may form separate bachelor herds, often with a dominance hierarchy, while females form herds which live in home ranges overlapping several male territories. This arrangement of male territoriality with female herds is found in some African ungulates. Males display to females in the rut so that there is a degree of sexual selection by females. Males of some species such as reedbuck (*Redunca arundinum*) and gerenuk (*Litocranius walleri*) defend their whole home range, while others such as blue wildebeest defend only a part of it, which may be referred to as a breeding territory.

Female herds are often matriarchal, as in the case of mountain goats. This arrangement is found in wildebeest, Uganda kob, asses (*Equus hemonius, E. africanus*) and Grevy's zebra (*E. grevyi*). The most extreme form of male display and territoriality is that of the lek. During the rut males display to females on a traditional breeding ground, where each male occupies a small area which it defends against other males. Some of these territories confer greater reproductive success on a male than others, while non-breeding males live in bachelor groups. Such a system has been observed in the hammer-headed bat (*Hypsignathus monstrosus*) and the Uganda Kob. The kob has a lek system with territories of 15-30m in diameter in some areas, whereas in others it has much larger territories of 100-150m diameter. In some places both systems are found. Leks are uncommon in mammals but are known in other classes of animals including birds and insects.

Spaced Males, Sociable Females

Males of some mammals are not territorial, but lead a relatively solitary existence, while females are more sociable. This situation is found in the coatimundi (*Nasua narica*), pigs such as warthog and elephants. Female elephants form matriarchal herds led by an old female. If a female is in oestrus, males are attracted to the female herd and compete for mating. Male African elephants, however, are more sociable than their Asiatic counterparts and 50% of African males live in loosely organized '*bull groups*'.

Multimale/Multifemale Groups

Multimale/multifemale groups are formed when adults of both sexes live together, permanently in a herd or troop. Although such groups are usually termed *'multimale'* by primatologists, to avoid confusion they will be referred to here as 'Multimale/multifemale' or *'bisexual'*. There is usually a male dominance hierarchy which may confer priority of access to oestrous females on high-ranking individuals. Females may also have a dominance hierarchy.

Burrowing mammals may live in Multimale/multifemale colonies as, for example, the black-tailed prairie dog studied by King and the European rabbit. The rabbit lives in groups of 1-3 males and 1-5 females, often with both male and female dominance hierarchies. Breeding success for both sexes is related to status, so that the dominant buck and doe consort. The colony defends the area around its home warren; dispersal is largely by males, 70% of whom leave their natal warren while 70% of females remain; female rabbits in a colony are there fore more closely related than males.

The multimale/multifemale herd is a mobile group in which there appear to be no personal relations apart from a male dominance hierarchy and the mother-offspring bond. This form of society was found by *Kaufmann* in the whiptailed wallaby (*Macropus parryi*), which appears to represent the peak of marsupial social organization. It is also characteristic of the African buffalo, which also forms all-male (bachelor) herds.

Female-bonded societies are the characteristic form taken by most higher primate bisexual groups. Males emigrate, while the females, which remain the their natal group, co-operate and show strong kin-related social ties with other females, often with a well marked kin-determined rank order. Males are hierarchically organized and may change troops several times in a lifetime. The best studied species are macaques (*Macaca fuscata, M. sinica, M. mulatta*). Two species of mammal are known which form non-female bonded societies. These are the chimpanzee and the closely related bonobo, or pygmy chimpanzee (*Pan paniscus*). The main social bond in these species is between males, and females leave their natal group to breed. Published data indicates that there is also a strong male-female bond in *P. paniscus*. These societies have a male hierarchy and *P. troglodytes* males tend to travel in bands, defending their group territory against other males.

Multimale/multifemale groups, known as prides, are characteristics of lions is some parts of their range. The social organization of the lion pride has been investigated by Bertram, Packer and Pusey and Schaller. In some ways it can be regarded as intermediate in form between a harem and a multimale/multifemale grouping. In this species, females form the stable social group; they are closely related (mean r = 1/8) and relatively uncompetitive, even suckling one another's cubs. Dispersal is by males which are driven out during adolescence. These males which are closely related (r = ¼) remain together and, on reaching adulthood, attempt to take over another pride from the resident males. Success in takeover is related to the number of males in the group, as is success in retaining it. Pairs of males or singletons have both a low rate of success in pride takeover and short tenure, so that they frequently join unrelated males to form a coalition of two or three individuals.

Table 4.5. The Frequency of Occurrence of Male Lion Groups of Different Sizes and the Probability that They will have Tenure of a Pride of Lionesses.

	No. of male groups			
No. of males in group	*Total*	*No. with tenure*	*Percentage with tenure*	*Significant differences*
1	23	4	17	$p<0.01$
2	39	23	59	$p<0.01$
3	12	11	92	
4-7	9	9	100	

Packer and *Pusey* estimated that approximately 72% of a male's partners were relatives. Between males of the same age or size there was no dominance hierarchy and, once a male consorted with an oestrous female, other males did not challenge him and ownership therefore conveyed temporary dominance. Fights occurred mainly when ownership of an oestrous female was uncertain. The degree of competition between males was strongly influenced by the number of females in the pride.

As all females tended to come into oestrous simultaneously, in a large pride every male was likely to find a consort simultaneously so that competition was minimal. Related males were no less competitive for oestrous females than unrelated ones. Female co-operation in rearing cubs is beneficial because the breeding success of males is dependent upon the presence of same age companions. Brown hyaenas (*Hyaena brunnea*) also form multimale/ multifemale groups with a rather unusual social system. *Mills* found that females only mate with nomadic males and no with male members of the pack or clan. Thus, the male has a choice of reproductive strategy, either to remain in his natal group as a helper or to become nomadic. Females are dominant to males and numerous females in the pack breed.

Eusocial

One of the most remarkable mammals from the viewpoint of its social organization is the naked mole rat (*Heterocephalus glaber*) which is eusocial with reproductive and non-reproductive castes, female worker being sterile. The details of its social structure have been reported by *Jarvis*.

Ephermeral Aggregates

Some mammals are sociable, spending much of their time in groups which represent ephemeral aggregations with no constancy of group membership. Herds of between one and 15 giraffes and eland are of this form. Solitary individuals are usually males; temporary groups may be of one or both sexes. Males are promiscuous, seeking out oestrous females for which they compete. Eland may also form nomadic female herds, while males tend to be more solitary, living in a small undefended home range; peer groups also form. Other reviews of mammalian social organization have been provided by *Eisenberg; Crook* and *Gartlan; Crook et.al.;* and, for African ungulates, *Leuthold*.

Territorial Behaviour

A territory is an area from which conspecifics are excluded by a combination of advertisement, threat or fighting. Territories may be defended by a single individual, a pair, a family or a social group or by individuals of either or both sexes. Males, however, are more commonly territorial than females. The size of mammalian territories varies; in some species, such as the rabbit, the female defends only her breeding burrow whereas in others the animal defends its whole home range. The shapes of territories are variable and usually relate to the distribution of resources within them. The European otter, for example, defends an elongated territory which follows the course of a river and provides both refuge and food for the animal.

Territoriality: Mammalian Examples

A good example of a mammal which defends its individual home range is the short-tailed shrew (*Blarina brevicauda*) studied by *Platt.* Using a radioisotope technique, *Platt* found that each shrew occupied a defended area of approximately 300m^2. Males and females held independent territories, although those of opposite sex overlapped in the breeding season. The boundaries were marked with secretions from the lateral and ventral glands. On meeting, shrews engaged in vocalization duels in which they screamed at one another and which sometimes led to physical assault and fighting. Most of the aggression which *Platt* recorded was between resident and nomadic shrews, not between territorial neighbours. Nomads were either the young of that year or residents emigrating from areas of low food density. Residents patrolled the boundaries of their territories but rarely intruded into hose of their neighbours.

The size of the shrew's territory was little lager than the area used daily but territory size was influenced by food resources, reflected in the density of field mice and invertebrates upon which the shrews preyed. As food supplies became less plentiful, so the territory increased in size until a critical point was reached when it became to poor to sustain the shrew, which then adopted a nomadic habit until it found a territory in a richer area. *Nomads* tended to remain on the boundaries of established territories, rarely intruding into occupied area. They vocalized only when in the process of displacing a resident from its territory, so that the vocalization appears to be a form of offensive threat. In the laboratory, shrews sometimes had vocalization duels which did not involve attack but simply led to the two animals dividing an area between them. In the majority of mammals individuals of the opposite sex are tolerant of one another with the result that territories are defended only against members of the same sex. Where male and female territories are of the same size with complete overlap, the species are monogamous. Examples are known from a variety of mammalian orders, including elephant shrews, klipspringers, coyotes and gibbons.

In this situation, each partner drives out members of its own sex. Where male territories are larger than those of females, a polygynous situation may exist if a male can totally include the territory of more than one female within his own. Male common tree shrews have been found to defend a territory which usually included that of a single female, but occasionally males had larger territories which included those of two or three females. The sportive lemur (*Lepilemur mustelinus*) is a territorial, tree-living, nocturnal mammal where both sexes hold territories. Male territories are larger (average 0.30 hectares) than those of females (0.18 hectares).

Defence is most marked in the male who uses surveillance posts in trees from which he can see his rivals, and both vocal and visual displays are given. The latter consist of shaking the head from side to side, leaping and shaking branches. Such a visual display is unusual in a nocturnal mammal. Neighbours display to one another and may even vocalize together, forming duets or trios. Male territories contain those of several females; up to five have been recorded and males appear to mate opportunistically with any female in heat. Frequently female mammals have an undefended home range, the selection of their living area being mainly depended upon suitable breeding sites such as burrows; males, on the other hand, are territorial and a male territory may contain several female home ranges.

Carl suggested that mate selection in a ground squirrel, the long-tailed souslik (*Spermophilus undulatus*), results from two independent factors, first the female's selection of a suitable burrow system, which she defends against other females, and secondly a male's dominance over a particular district. To obtain maximum mating success a male should acquire a territory which includes several suitable sites in which females can breed. In the species so far described, non-territorial individuals are nomadic, avoiding occupied territories unless challenging the owner. In some species, however, non-territorial males may be tolerated and occupy subordinate status within a territory. *Bearder* and *Martin* found this situation in the lesser bushbaby (*Galago senegalensis mohohi*) where a territorial male associated with females and young, while subordinate and juvenile males lives peripherally and did not breed. A similar hierarchical system also operates in prairie dogs, studied by *King.*

Each social group or 'coterie' is confined to a territory held by a dominant male, who displays by leaping and rearing up on the hind legs while uttering a trisyllabic call. The coterie boundary is traditional and may be maintained over many generations even though the territorial males may change. Females defend their breeding burrow when rearing families. The young learn the boundaries of the coterie and males tend to emigrate to take over other coteries, while females remain in the area of their birth. The male white rhinoceros is territorial and his breeding success depends on holding territory. Two-thirds of adult males (alphas) hold territories, while females are more mobile and move through the males' territories. A territorial male actively attempts to prevent an oestrous female from leaving his territory by standing between her and the border. Alpha males tolerate subordinate males (betas) which live in their territory and, on meeting, the beta emits a submissive shriek or snarl. Unfamiliar males are attacked. Alpha males meet very occasionally at the territorial border, on average once every two weeks.

Sometimes the two males charge and wrestle horn-to-horn. Trespassing alpha males, often on the way to a water hole, five the submissive snarl and retreat on meeting the territory owner. Alpha males mark the territory with dung, which the animal kicks and spreads out, and also spray urine. Beta males and females defecate on dung heaps approximately half of the time but do not spray urine. Dung heaps are especially prominent on territorial boundaries are spray-urinating by alpha males occurs both on the boundary and on well-marked trails. Territorial takeover occurs when a male is successfully challenged by a beta male from another area. The former alpha is tolerate and remains in his territorial area but ceases to kick dung and spray-urinate. The new alpha male takes over the old territorial boundaries but also attempts to increase his territory size. Juvenile white rhinoceros are driven away by the mother at 2-3 years of age on the birth of a new offspring, when they join subadult groups of up to six individuals. Cows separate from these groups at 9-10 years old and bulls at 11-12, when they become fully-grown and sexually mature.

In more social species it is common for non-breeding males to join bachelor herds. Breeding males may hold year-round territories, with females moving into these areas during the rut, but in other cases males are only territorial during the breeding season. Sometimes males defend very small areas which are used solely for breeding, a situation which is common in eared seals (Otariidae) and some true seals (Phocidae). Male Steller sea lions defend an area of the beach while includes a group of females; as the females are relatively mobile, individuals may come and go at

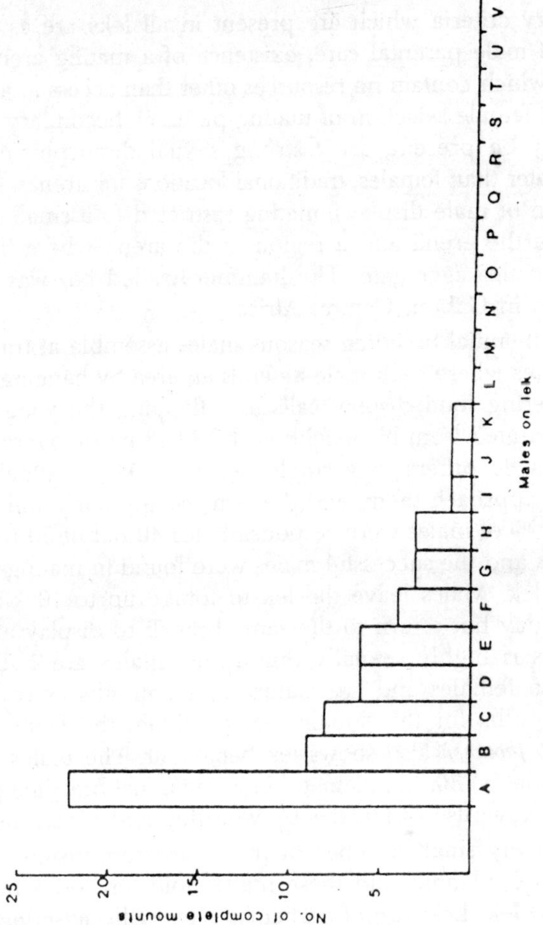

Fig. 4.14. Steller sea lion (Eumetopias jubata) territorial behaviour and courtship.

will. A similar situation was recorded from the grey seal by *Fogden*. Weddell's seals (*Leptonychotes wedelli*) breed on ice and the females congregate near an ice hole around which a male defends an underwater territory. Males (unusually for a polygynous species) are actually smaller than females. Another example of male territorial behaviour restricted to the breeding season is that of the lek. Females appear to mate selectively with males holding certain territories within the lek. *Bradbury* laid down certain criteria for defining a lek breeding system which he divided into primary and secondary features.

Primary criteria which are present in all leks are as follows: absence of male parental care, existence of a mating arena, male territories which contain no resources other than access to a female, and finally female selection of mating partners. Secondary criteria, which may be present, are : strong sexual dimorphism, males maturing later than females, traditional locations for arenas, extreme ritualization of male displays, mating restricted to a small number of males in the arena and a region in the arena where the most successful males aggregate. The hammer-headed bat was studied by *Bradbury* in Gabon, Central Africa.

In the biennial breeding seasons males assemble at traditional breeding sites where each male defends an area by hanging from a branch uttering loud display calls and flapping the wings. Each male is separated from his neighbour by 13-18 m, on average with as many as 84 others in a single lek area. Males attack other males who approach them, and the females approach and inspect the males; 6% of males were responsible for 40 out of 50 recorded copulation's and the successful males were found in mating centers within the lek. Males leave the lek to forage up to 10 km away during the day but return to the same branch to display at night. The species is highly sexually dimorphic; males are 1.79 times heavier than females and are mature at 18 months as compared with six months for the female. An antelope, the Uganda kob (*Adenota kob thomasi*) also shows lek behaviour. The males defend small territories 15-30m in diameter in a traditional breeding ground where they advertise to females by whistling and visual display.

A relatively small number of males are responsible for the majority of copulations and these males hold territories near the center of the lek. Leks are of particular interest because territories apparently contain no particular resource. It has generally been asserted that males compete for particular locations but *Bradbury and Gibson* cast doubt on this and considered that the criteria for female choice are still unidentified. Some gregarious mammals defend group territories. Both sexes co-operate for defence purpose in wolves, hyaenas and common marmosets, but defence is the prerogative of the male in lions and chimpanzees. Matrilineal groups of common langurs defend territories, while a male simply defends the group of females against rival males.

Social Behaviour among Wolves

Wolves are large members of the genus *Canis*, closely related to domestic dogs. Adult males average about 43 to 45 kilograms

(95 to 100 pounds, about the size of a good-sized German Shepherd). A few individuals may weigh more; an 80 kilogram male was reported from Alaska. Wolves are carnivorous and predatory; their usual prey consists of large animals, such as moose caribou, and deer, although they sometimes eat smaller herbivores, such as rabbits and mice. Their only predators are human but they are host to many parasites, such as ticks and tapeworms, and disease, including rabies. The habitat of the wolf once included all of the biomes of the Northern Hemisphere, excluding only tropical rain forests and deserts.

At present, the only substantial numbers of wolves in North America are found in Canada (the estimated number there is 17,000 to 28,000) and Alaska (5,000 to 10,000). The only population left in the rest of the United States, in northern Minnesota, numbers about 500 to 1,000 individuals. The state of Wisconsin placed the wolf under legal protection in 1957, but that was too late; there are no more wolves in Wisconsin. An additional 20 to 30 wolves live on Isle Royale in Lake Superior. There also a few wolves left in upper Michigan, and there may be a small population in the Rocky Mountains. Now the Endangered Species Act of 1973 provides legal protection for all the wolves in the 48 contiguous states.

The Wolf Pack

Carnivorous animals that prey on large mammals either weigh as much as there prey or hunt in packs; wolves belong to the latter category. Wolf packs are usually made up of two to either members, although much large groups have been reported occasionally, as have solitary wolves. A pack in this size range seems to represent a maximally efficient hunting unit; *L. David Mech*, author of what is considered the definitive study of wolves, reports that only about six wolves of a large pack of 15 actually hunt at any one time. Natural selection would also work against tendencies to form very large packs since a single kill would not feed all members of such a pack. However, since packs very greatly in size–as compared to clutch size in birds, for instance–selection pressures must also be either varied or not very strong. Although a number of field studies of wolves have been done, on one has been able to observe a single wild wolf pack over a long enough period of time to determine how a pack originates. Most packs are made up of a breeding pair, pups, and extra adults old enough to breed (but which usually do not). The social bonds among members of a wolf pack are very strong.

In both strength and expression they closely resemble the bonds between a dog and its master, and, indeed, it seems certain that the existence of a strong bonding system among the evolutionary ancestors of the domestic dog is the reason why the dog is so uniquely domesticable. Wolf packs travel, sleep, and hunt together. They share their kills, either by regurgitating food, by carrying home part of their prey, or by leading other members of the pack to a fresh kill. This sharing of food among adult members of a group is rare among wild animals; for example, among the primates, organized sharing of food occurs only in our own species. Studies made of animals in captivity indicate a strong inhibition among wolves against taking food away from an animal already n possession of it, even by wolves that are both superior in strength and hungry. One of the strong bonds that holds the wolf pack together is the sexual bond between the breeding pair. Wolves may begin choosing mats when they are about one year old, although a wolf is not sexually mature until the age of 22 months, and there is some evidence that these pair bonds, once formed, are maintained until one member of the pair dies.

The bond between a pair of wolves is not related to the sexual availability of the female. Female wolves come into heat only one a year, and although courtship (which closely resembles the play behaviour of younger animals) is carried out over a period of several weeks, copulation occurs only during the one week of estrus, the time when ovulation takes place. It has been suggested that this bonding is reinforced by the copulatory tie, a phenomenon unique to members of the dog family. In the copulatory tie, the female's vaginal sphincter muscles constrict about the swollen penis of the male after intromission, and, as a consequence, the two animals remain attached by their genitals for 15 minutes or more following ejaculation, during which time they stand or lie rump to rump.

The copulatory tie immobilizes both animals in an extremely awkward position, highly vulnerable to harassment and attack. To the casual observer it would to seen likely to increase the affection of the individuals involved toward one another. Recent studies, however, indicate that it may increase fertility. Strong bonds are formed between mothers and pups and between the pups and other members of the pack. During the pups' first three weeks of life, they remain in the den, in close contact with one another, and are

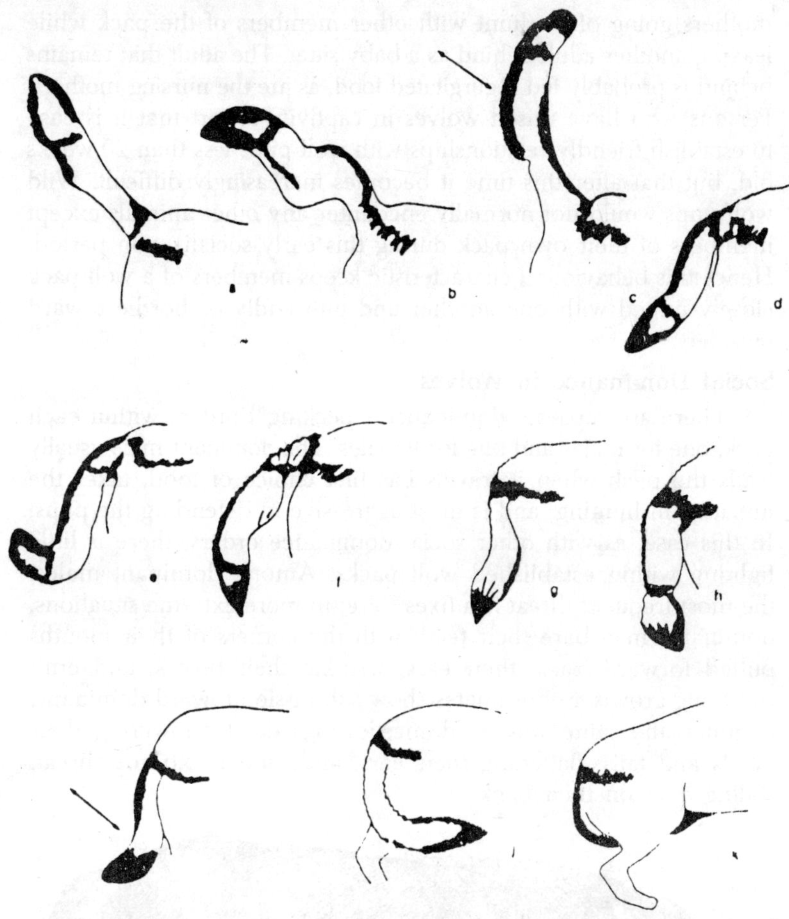

Fig. 4.15. Tail positions and their contexts in the wolf (Canis lupus): arrows indicate lateral movement. (a), (b) confident, in locomotion, (c) confident, threatening, (d) relaxed position, (e) slightly apprehensive threat, (f) relaxed position, usually eating or watching, (g) tense, (h) defensive threat, (i) submissive, (j), (k) extremely submissive.

nursed by their mother. After this time, they begin to et predigested food that they obtain by touching their mouths to the mouths of adult wolves, which stimulates the adults to regurgitate meat for the pups. Wolf pups are fed in this manner not only by their mother but by any adult member of the pack.

As the pups grow older, they are played with by all members of the pack, and numerous instances have been reported of wolf mothers going off to hunt with other members of the pack while leaving another adult behind as a baby sitter. The adult that remains behind is probably fed regurgitated food, as are the nursing mothers. Persons who have raised wolves in captivity report that it is easy to establish friendly relationships with wolf pups less than 20 weeks old, but that after this time it becomes increasingly difficult. Wild wolf pups would not normally encounter any other animals except members of their own pack during this early socialization period. Hence this behavioural characteristic keeps members of a wolf pack closely united with one another and unfriendly or hostile toward outsiders.

Social Dominance in Wolves

There are separate dominance ("pecking") orders within each pack, one for males and one for females. The dominant male usually leads the pack when it travels has first choice of food, takes the initiative in hunting, and is most aggressive in defending the pups. In this case, as with other social dominance orders, there is little fighting within established wolf packs. Among dominant males, the most frequent threat is a fixes stare; in more extreme situations, dominants may bare their teeth with the corners of their mouths pulled forward, raise their ears, wrinkle their brows, and emit rumbling growls. Subordinates show submission toward dominants in much the same way as domestic dogs do, by lowering their heads and tails, flattening their ears, and, under extreme threat, rolling over on their backs.

Fig. 4.16. Canis lupus (Wolf).

Subordinate members also often express affection for the dominant wolf, nuzzling the mouth of the pack leader, using the food-begging gesture of the pups. Sometimes all the members of a pack will surround the leader and lick his face and poke his mouth with their muzzles. This ceremony often occurs when wolves first wake up, when the group has been separated from one another for some time, and when the pack has scented prey and is ready to start on a hunt. Usually the dominant male ad the dominant female are the only mating pair. Since all the male members of the pack prefer the dominant female, and all the female members prefer the dominant male, courtship usually does not occur between other pack members. If it does, it is disrupted by the dominant animals. Hence social dominance has the effect of limiting the size of the group.

According to some studies, as many as 40 per cent of all adult female wolves do not breed in any given year, and fewer than half–in some studies as few as 6 per cent–of the pups survive for one year. When packs are heavily hunted, the ratio of pups to adults increases, indicating that more pups are produced or survive, or both, under these conditions. In short, wolf packs tend to remain stable in size. When the dominant male of Female dies, subordinate wolves will then mate and bear pups. It has been suggested that new packs are formed at times of social unrest after a leader has died, leaving two adult males of almost equal rank.

Territoriality

Wolf packs are territorial each pack occupying a home range of 130 square kilometers or more. A wolf pack intruding on another's home range is attacked aggressively. Two adjacent ranges may overlap, but since the areas of overlap are small and the territories very large, the chance of two packs encountering one another is not great. The boundaries of territories are scent marked with urine; scent marking probably conveys information about both the extent of the territory and the amount of time that has passes since the pack passed by thus making it easy for packs to avoid one another. Howling may also be related to territoriality, analogous to singing in male birds. When a pack howls, one wolf begins and the others join in, finally working up to full chorus. A session lasts about a minute and a half and is obviously pleasurable for the wolves, as are howling sessions for domestic dogs–somewhat like a community sing.

Howling can be detected by the human ear up to 6 kilometers away, and probably for a greater distance by other wolves. The howl of each wolf in the pack is different. One observer reported, in fact, that if he howled on the same note as his pet wolf, she would shift her howl by a few notes. Thus a howl could provide information to a neighbouring pack not only of the presence of another wolf pack but also of the number of members in it. Howling also seems to serve as a way to regroup pack members that may have been separated during a hunt.

The Wolf as Predator

Wolves prey mostly on large animals, including deer, moose, caribou, elk, and bighorn sheep. Characteristically, they consume almost all of their prey, including hair, hide, and all but the largest bones. If the entire carcass cannot be eaten on the spot, the remains are either carried away or revisited. Wolves often locate prey by scent, sometimes from a distance of most than 11/2 kilometers. Staying downwind, they first stalk the animal slowly. They begin to run only when the quarry runs; apparently the movement of the prey animal serves as a stimulus. The beginning of the chase is the crucial part of the hunt; this is actually a testing period.

Generally, the only prey animals that are chased for more than a few minutes are those that falter because of immaturity, old age, or illness, or calves that have been deserted by their mothers. Healthy adult animals of all prey species can easily outrun wolves. Larger animals often simply stand their ground; a full-grown moose can weigh over 450 kilograms and is well able to kill a wolf. Domestic animals such as sheep and cattle have no such advantages, however. Having evolved under human protection they are typically unable to defend themselves either by fight or flight and so are an easy target for a wolf pack or even for a lone animal. There have bee occasional reports of wolves using cunning and complex strategies in their hunts–such as driving the prey to a predetermined ambush–but if such tactics occur at all, they would appear to be rare. It is difficult to state with certainty whether predation by wolves serves as a form of population control for prey species.

The principal limitation on the population size of large herbivores appears to be the food supply, yet, as we noted previously only rarely does a population under natural conditions expand to the point where large numbers of individuals die from starvation. Therefore, it can be argued that they prey population remains at

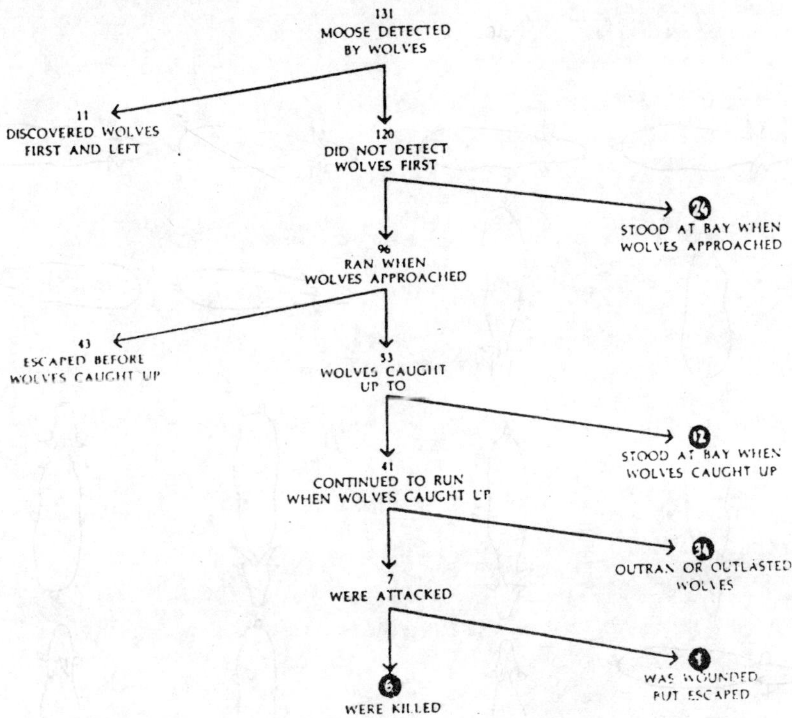

Fig. 4.17. Results of interactions between all or part of a large wolf pack (15 to 16 wolves) and 131 moose on Isle Royale.

about the size that resources can support, whether or not wolves are present. (In this regard; it is interesting that before wolves arrived on Isle Royale, almost on twin calves were ever seen among the moose population; by 1963, the twinning rate had reached 38 per cent). Other contend that the "culling" of the population by wolves benefits the prey, not only by removing ill and unfit individuals but also by holding down the population. In Yellowstone National Park, where wolves have been exterminated, great herds of elk have ravaged the vegetation and starve periodically.

Clearly wolves and their prey can coexist for long periods of time, and regardless of the presence or absence of wolves, populations of large herbivores cannot expand indefinitely. Moreover, the very young, the old, the ill, and the inept will be the least likely to survive against almost any selective force. The wolf is in the unfortunate position of being in direct competition with humans for the position of top-level carnivore. Every place the competition

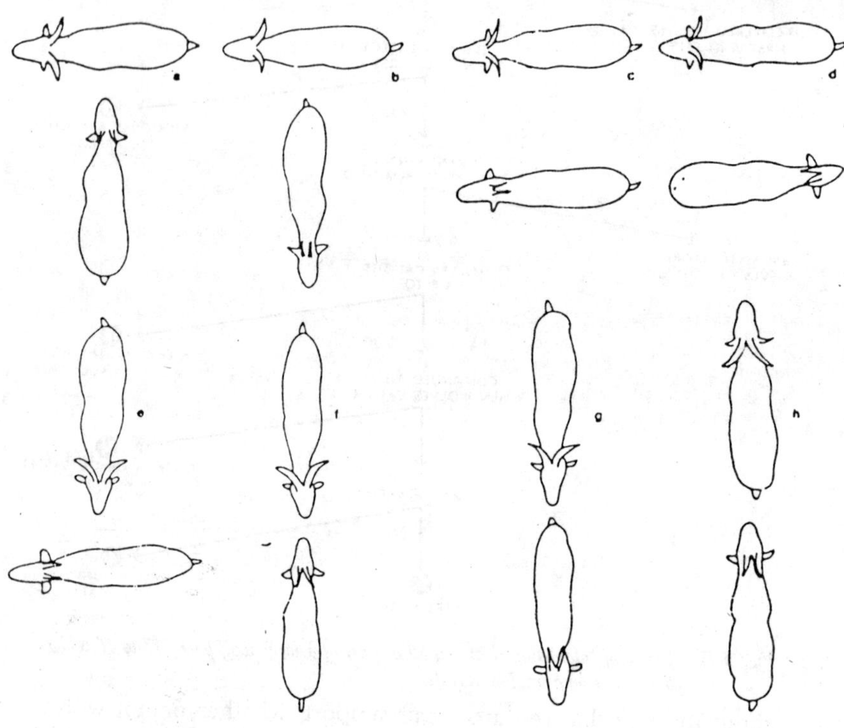

Fig. 4.18. *Directional signalling by a territorial male Grant's gazelle to a female of his harem, male with larger horns. (a) Stop! do not advance in this direction! (b) Do not turn round! (c) Stop or go ahead! (d) Do not turn round! (e) Continue in your direction and speed up! (f) Turn round and withdraw! (g) Go ahead! Follow me!*

has been intense, as throughout the contiguous United States, the wolf has been exterminated, and, with the new hunting methods available the problem of wiping out the remaining relatively small wolf populations in Alaska and Canada is not a difficult one should humans, pursuing Gause's principle choose to do so.

Social Organization Among Baboons

Baboons are large, quadrupedal African monkeys with protruding, dog like muzzles. One genus (*Papio*) and four species are commonly recognized: a savanna species (*Papio anubis*), of which there are at least two, and probably more distinct races; two forest species (the drill and the mandrill), both short tailed baboons living in West Africa; and a desert species, found in North Africa and Arabia, *Papio hamadryas*, the hamadryas baboon. The baboons are a particular ecological interest for two reasons. First, unlike most monkeys, they live on the ground and have done so for several million years. For this reason, differences between baboons and other monkeys can be related to some degree to their environments.

Second, there is a marked difference in social organization between the savanna and forest species, on the one hand, and the desert species, on the other. Since the groups are closely related genetically, obviously sharing a common ancestor in the not-too-distant past, it is very likely that the differences can be explained ecologically rather than genetically. Both of these statements must be tentative, however; field studies of baboons have been made only since the late 1950s, and most other monkey groups have been studied inadequately or not at all.

The Genus Papio

Baboons are large, with adult males weighing about 55 kilograms. Their size is clearly related to a terrestrial existence; among the primates, only large animals, such as baboons, some of the great apes, and *Homo sapiens*, are terrestrial. Among all baboons, adult males are very different in appearance from adult females. This phenomenon, known as sexual dimorphism, is much less pronounced among other primates and particularly other monkeys. The males have mantles and prominent canines and, most obviously, they are about twice the size of the females. Sexual dimorphism, which is much less pronounced among tree-dwelling primates, appears to be related to the increased hazards of life on the ground and to the male role of defender. Like most primates, baboons are almost exclusively vegetarian, living on leaves, fruits, flowers, and young stems, all of which they pick and consume on the spot. They also dig for roots, bulbs and tubers. This vegetable diet is supplemented by insects and occasionally by meat, such as snakes, lizards, fledgling birds, or young gazelles, when they are encountered accidentally.

Fig. 4.19. Semnopithecus (Baboon).

There is no organized hunting for prey animals and no sharing of prey. Baboons sleep in tall trees or on cliffs, returning to a nesting site every night. Hence, they must often travel to find food, sometimes for great distance across open country. Baboons have a harder time finding food than do tree-dwelling primates, which are more likely to live in the lush tropical rain forests and which sleep where they eat. The difference in size between the two sexes of baboons is probably related to the difficulties of the animals in finding food. The male is large enough to act as protector, whereas the female's smaller size reduces the food requirements of the family unit and so promotes survival of the young.

The Structure of the Band

Baboons are organized into multimale bands; those observed have ranged in size from eight to more than 185 individuals. Adult females outnumber adult males by about 2 to 1. Much of this difference can be accounted for by the much slower maturation rate of the males. Although an adult male can probably produce viable sperm by the time he is about four years old, he does not reach full size nor are his canines fully erupted until he is eight or more, and therefore is not counted as an adult male either in the social organization of the band or by the field observer peering through binoculars. Females, on the other hand, are adults by the age of two or three. The bands are territorial, with fairly distinct home ranges, and bands rarely meet. When they do, the one farthest from the center of its customary range generally moves away, sometimes after an exchange of vocalizations, canine displays, and gestures, but more often with no visible reaction at all, except perhaps a slight display of nervousness. Bands that know each other may share large sleeping groves.

Overt fighting has been reported once; it occurred between two groups trying to sleep in the same clump of trees. When the bands move in open country, females and infants characteristically are in the middle of the group, close to adult males, Mass counterattacks by adult males, who placed themselves between the group and the attackers, have been observed to route a leopard and disperse a large dog pack, killing or wounding several of its members. Another function of the band is the pooling of information.

Fig. 4.20. Ibex (Capra ibex) threatening a rival by standing on its hind legs.

Hans Kummer, one of the most experienced and imaginative observers of baboons, describes a band at the beginning of a day's much as resembling a giant amoeba, stretching out and then withdrawing one pseudopod after another, as small segments begin to move off in one direction, only to rejoin the band if the others do not follow. Finally, a movement is initiated that meets with general approval, and the day's much begins. The size of the group seems to be somewhat related to its habitat. Very small groups are correlated with small areas of vegetation, quite widely separated from one another, whereas larger groups may be found where a typical feeding site might be a clump of trees. Logically, troop size must be a compromise between the availability of food resources and the need of the troop to defend itself.

Social Dominance

In all baboon species except the hamadryas, the band is organized around a dominance hierarchy of adult males. As in other dominance hierarchies, the band is maintained mostly by behavioural conventions, with only occasional fights, which rarely result in injury. Dominant males threaten subordinates by staring, yawning (with the ears laid back), raising the eyebrows, and, at close range, by grinding their teeth. Subordinate males demonstrate submissiveness by looking away, grinning, turning their backs, and presenting their hindquarters. The dominant male may mount the presenting male briefly. The gesture of presentation, also used by females in estrus as an invitation to copulate, is the most reliable index of hierarchical position. The top-ranking male presents to no other individual, the second-ranking male only to the top one, and so on down the social ladder. (Presentation as a gesture of submission or appeasement is seen among many primate species).

Adult females are subordinate to males but also have a separate dominance hierarchy, although not as clearly structured as that of the males. Adult females present to superior females. Socially superior males mate more often than inferior males. In some groups, it was observed that a number of males copulated with females when they were not in full estrus, when the likelihood of fertilization is considerably less. But when a superior female is in full estrus, as revealed by maximum swelling and colouring of the genital area, only the highest ranking male mates with her. Thus, the highest ranking male is the one most likely to father offspring.

Dominant males also have first choice of nesting sites. However, because there is no shared food supply, dominants are not allocated a larger share of food than subordinates, as they are in some other pecking orders. Dominant male play a leading role in territorial disputes, thus maintaining the spacing among bands. They are also the most aggressive in defending the band against intraspecific aggression. If another member of the group bark a warning or screams in fear, the dominant male steps forth and inspects the danger.

Social Bonds

As among almost all higher primates, grooming is a prominent from of social behaviour among baboons. It involves a brisk parting of the fur of another individual with the fingers of both hands and the picking off of small particles, including insects; these items are often swallowed. Dominant males, females in estrus, and females with infants are groomed most frequently, but all members of the group receive some grooming attention. The function of grooming is clearly not only hygenic but also a continual reinforcement of social bonds. Infants are another clear social bond. The infant, which is all back and therefore very conspicuous for its first few weeks of life (perhaps so it can be guarded more zealously), usually clings to its mother, who may support it with her arm. With the mother's permission, it may be touched or even held briefly by other members of the troop.

The infant or infant-mother pair are clearly attractive to other members of the group, who tend to cluster around, especially when the infant is newborn. Mothers with infants are groomed frequently, particularly by other females, and dominant males attend them closely as the troop moves. An infant that loses its manner will be adopted, sometimes by a childless female, but often by a young male.

The Hamadryas Baboon

The hamadryas baboon is unlike the other members of the genus *Papio* in that there are three distinct levels of social organization. The principal unit is a male with one or more females and their young, totaling up to seven or eight members. These one-male units group together into bands, similar in size and organization to the bands of the other baboon species. Finally, the bands come together in troops, some of which have been counted as containing 750 members. Students of primate behaviour hypothesize a direct relationship between these social organizations and the environmental resources of the hamadryas.

The habitat, which is one he edges of deserts, is typically arid grassland, interspersed with thorny acacias and other small trees and bushes. There are no tall trees, and the baboon of this area sleep on ledges on the vertical slopes of sleep cliffs. (This is learned behaviour. The hamadryas will sleep in tall trees if they are available, and other species will sleep on cliffs in there are no tall trees.) The baboons move across open country, often for long distances, between their sleeping cliffs and their feeding sites. They travel in bands, break up into one-male units for feeding–typically one unit to a tree–regroup into bands, return to the cliffs, where they often sleep in troops, and then recongregate each morning in the band. Thus each social unit serves a clear and important function.

The one-male unit is an optimal foraging unit, provided with a protector; the band provides for mutual defense when travelling; and the troop makes possible maximum utilization of safe sleeping sites within the habitat. The strongest bonds are those that hold the one-male unit together. The female mate exclusively with the unit leader, and almost all social interactions, such as grooming, are carried out within the unit. Juveniles sometimes leave to play with those of other units within the band. The structure of the hamadryas band is generally similar to that of other baboon species. There is a male dominance hierarchy, and males act as defenders.

About 20 per cent of the adult males are not part of any family unit, but, as with other baboon species, these males are part of the band. The band moves as a whole, apparently acting on information from knowledgeable members of the group. The ties that hold the band together are less strong than those that hold the one-male unit of the hamadryas together. If a member, because of age, illness, or injury, drops out of a nonhamadryas band, he or she is left behind. If a disabled female drops out of a hamadryas band, her male leaves the band to stay with her. The band, however, like the one-male unit, is a stable social group, always composed of the same members. The troop, on the other hand, is made up of varying groups of bands from the same territory that apparently recognize one another and are mutually tolerant. Because the male-dominated band is the primary social structure of the other *Papio* species (and also of the macaques, arboreal monkeys who are their closest relatives), the hamadryas band appears to be a legacy from their common ancestors.

Similarly, the troop seems easy to understand as a direct outgrowth of environmental pressures; baboons of other species share water holes and large sleeping groves when they have to. What, however, are the bonds that hold the one-male unit together? The cohesion of the unit turns out to depend primarily on two behavioural patterns, both those of the male. The first is a herding instinct toward the females. The females are taught to follow the male. He constantly watches them over his shoulder and, if one drops back or attempts to slip away, he stares, threatens, and sometimes nips her on the back of the neck, aster which she immediately follows. If a male hamadryas is presented with a female olive baboon (*Papio anubis*), he accepts her quite readily and immediately trains her to follow him. The second behavioural pattern is a strong inhibition baboon males against taking another male's females. This, too, was tested experimentally. Two adult males, who knew each other, and a strange adult female were trapped. One male was enclosed with the female while the other was permitted to watch from a cage 10 meters away.

The male caged with the female immediately made grooming, mounting, and herding advances towards the female, as he would toward a member of his harem. Fifteen minutes later, the second male was put in the same enclosure with the pair. He not only refrained from fighting over the female, but he avoided even looking at them. The animals were separated, and two days later the experiment was repeated again, this time switching the males. Once again the male that had watched the activities between the male-female pair was strongly inhibited in his behaviour. Another adaptive feature of this inhibition is that it applies only to adult males and females.

Young males are attracted to juvenile females, which they "kidnap" from their family groups before they are sexually mature and train to follow them. Thus, the younger male has a detour around the inhibition barrier, and a new family units can come into existence. Lest you be tempted to extrapolate to human behaviour from these accounts of baboon behaviour, it is of some interest that, among the baboonlike geladas, groups are formed and held together by activities of both sexes. Dominant males pair with dominant females. Wife number one is dominant over wife number two, and so on. The geladas are large monkeys that live in the mountainous grassland of Ethiopia. Although they are referred to as baboons in common usage, they are different enough to be recognized as a separate genus (*Theropithecus gelada*). The baboons remind us that social behaviour in animals is ecologically adaptive–that is, that it promotes the survival of members of the social organization (and particularly of their young) within a given habitat or range of environmental conditions.

5

SOCIAL BEHAVIOUR IN WOMEN

Much of the past ethnographic and theoretical literature on humans implies that females are subordinate, passive, and dependent on males, and that their strategies have developed in response to male strategies and needs. Although numerous researchers have recently pointed out the fallacies in such statements and Lamphere, 1974. Reiter, 1975; Iglitzin and, Ross, 1976, Jacobs, 1981), relatively few authors have addressed the potential consequences of these value judgments for people presently experiencing rapid westernization. The intention of this chapter is to identify some of the sources and possible effects of such male-focused value judgments (see Wasser and Waterhouse, this volume) on women's health care in the patrilocal-patrilineal system of the Taiwanese.

A. Stereotypic Images of Women

As is typical in patrilocal-patrilineal cultures, a Taiwanese woman usually lives with her husband and his family after marriage, and inheritance is passed along the father's lineage. This mode of inheritance and residence is also characterized by the existence of different social strategies and sex-roles between women and men that are vulnerable to misinterpretation (Wolf, 1972). Wolf (1974) cites several examples of male-focused stereotypes regarding women in such patrilocal societies:

1. Female children are said to be less desirable than male children.
2. Women are also said to impart little sociopolitical influence.
3. Women and their roles are often seen as secondary to those of men.
4. Women are reported to participate in meaningless social groups.

However, alternative views of women in this culture have been provided by Wolf (1972, 1974, 1975), Ahern (1975), and Harrell (1981a, 1981b). The work of these authors reveals that such stereotypes often reflect the use of value judgments by researchers when evaluating both the sexual division of labour in Taiwan and the ways in which Taiwanese women cope with their particular residence system (see Section II). Unfortunately, many of the male-focused values continue to persist today, and their negative consequences for the Taiwanese health-care system, as it undergoes westernization, are now becoming readily apparent, as explained below.

B. Sex-Role Differences, Traditional Values, And Health Care

Indigenous medical systems evolve according to the ways in which societies are organized (Worth, 1975; Kunstadter, 1975) and in response to the perceived needs of the people (Kleinman and Kunstadter, 1978). As a result, they tend to meet the psychosocial and medical needs of those individuals who adhere to traditional beliefs (Gallin, 1978). Prior to the introduction of Western medicine in the nineteenth century (Gale, 1978), both women and men presumably employed traditional healing methods to similar degrees. However, Taiwanese women are currently twice as likely as men to use traditional methods of health care (Tseng, 1975; Kleinman, 1980). The sex differences in health-seeking behaviour appears to result, in part, from sex differences in adherence to traditional sex-roles, and in differences in degrees of exposure to westernization. Taiwanese men tend to have a relatively broader range of outside enterprises available to them than do women (Wolf, 1975), thereby exposing them to a variety of Western ideals, medical systems, and value judgments. This is especially the case for men who work in wage labour.

Women, on the other hand, typically remain more family-oriented, even when employed outside of the home, because Taiwanese philosophies hold women responsible for family cohesion. The maintenance of family cohesion is accomplished, in part, through adherence to traditional concepts and values, including traditional medical treatments. However, the amount of information about and accessibility to health-care methods are commonly dependent on an individuals age, sex, status, and the value placed on his or her work (Odum, 1971). As a result, the relatively greater westernization of men, combined with Western ideologies that promote a higher value on male activities, may be depleting the availability and credibility of the traditional health-care methods. The availability of traditional medicine in Taiwan is further reduced because many social scientists and government officials, who organize and operate medical licensing and health services, tend to promote Western medical programs over traditional programs (Gallin, 1978). This loss could be particularly disruptive for many Taiwanese women and others who are less westernized, since traditional healing is often more culturally appropriate for them. In this chapter, I first address Wolf's (1974) four points regarding the subordinate images of Taiwanese women (cited in Section 1, A) to illustrate why previously applied value judgments and portrayals of women have been misleading. I then show how life-stage and role differences between the sexes, combined with the accessibility and appropriateness of medical systems, contribute to sex differences in health seeking behaviours.

Finally, I discuss how the influx of Western values and medicine, in the presence of an erroneously male focused value system, has contributed to a dualistic health-care system that is becoming increasingly dominated by western medicine. The discussion will emphasize the unfortunate consequences that could result from the loss of traditional medical systems-systems that could benefit many Taiwanese people when properly used in conjunction with the more technologically advanced medicine of Western cultures.

The Environment, Division of Labour, and Sex Roles

A. Ch'i-nan's Complementary Sexual Division of labour

An ample boy of work has been done on the village of Ch'i-nan, Taiwan. This village is discussed here as a representative "model" of how a sexual division of labour that is beneficial to both sexes has developed in that area of Taiwan. Sex roles are quite complementary and thus, female roles are not seen as secondary to those of men. When the inhabitants of Ch'i-nan immigrated from mainland China to Taiwan approximately 200 years ago (Harrell, 1981a), they were already rice farmers engaging in a distinct division of labor and patrilocal residence.

The environmental conditions in Taiwan favoured the continuation of those previously established characteristics. However, farmable land was scattered in the Ch'i-nan area, thereby promoting the formation of discrete social units of related individuals for its cultivation and inheritance (Harrell, 1981). At the same time, other environmental constraints promoted cooperative efforts between these separate social units. For example, invading neighbouring settlements often forced the individual agnatic units of ch'i-nan to unite for defensive measures (Harrell, 1981a). Intergroup cooperation was also facilitated by the establishment and maintenance of communal irrigation systems, which were critical to the development and production of wet rice farming. The patrilocal residence system, which assures a stable community of male members, was particularly advantageous under these conditions.

B. The Allocation of Sex Roles

In addition to the activities above, other needs had to be fulfilled. These included (a) health-care maintenance, (b) socialization of young and (c) the production and processing of other foods critical to a balanced diet. However, the performance of these latter, tasks was not always compatible with wet rice fanning, defense, and irrigation activities. For example, working in, and sometimes defending, the water-soaked rice fields would present obvious difficulties for nursing mothers and their infants. Thus, a sexual division of labour, whereby all needs could be met, would seem most efficient and hence beneficial for both women and men. As a result, men worked in the fields, while women were responsible for a variety of tasks that were more compatible with child-rearing and maintaining family cohesion (Wolf, 1975). Moreover, family cohesion appears to be critically important in the Taiwanese culture.

A Women's Life-History Stages

Given the preceding sex differences in roles and natal dispersal patterns, one might suspect that child-rearing practices would differ according to the sex of the child and the ways each sex copes with the environment at each life stage. In fact, as will be shown, some of the major reasons for the male-focused stereotypes in this culture cited by Wolf (Section 1, A) appear to result from a failure by Westerners describing this culture to take the need for these sex differences into consideration. For example, young girls contribute to many family needs while they are simultaneously being prepared for the patrilocal residential transition.

Since the eventual transition results in a complete alteration of social and environmental resources available to women, young girls should be developing a heightened awareness of social and environmental conditions during this first stage (Wolf, 1972). Females then use the learned domestic and, social skills to develop appropriate strategies in coping with the new social and environmental conditions. In essence, each life stage prepares them for the next. In their third life stage, women utilize the power, influence, and knowledge that they have accumulated through interpersonal relationships with family and social group members.

Section III details each of the three life stages of women: daughter, mother, and grandmother, to illustrate how women cope with their particular social system. This section also shows how women's health-care strategies are interwoven with their roles– women respond to family illness in accordance with their biological, environmental, and cultural needs, and according to the health-care methods that are available to them.

A. Stage One The Daughter

Patrilocal residence and adult role expectations in Taiwan require a more demanding socialization process for daughters than for sons. For example, daughters are expected to perform many more domestic activities at an earlier age than sons (Wolf, 1975). Furthermore, the performance of domestic responsibilities is supplemented with the necessary fulfillment of a girls own developmental needs (e.g., forming peer relationships). The structured and seemingly restrictive norm for daughters could be perceived as an indicator of female subordination. However, the responsibilities of familial and personal activities seem to provide a young girl with valuable skills and abilities (e.g., organizational skills patience and creativity, and an overall heightened awareness of her physical and social environment) that will eventually enhance her success in the transition to adulthood.

The acquisition of such abilities requires that young girls explore diverse behavioural options and social techniques (Wolf, 1972), which in essence provide her with a foundation of diverse and broad-ranging skills to use in response to life changes. Another potential indicator of female subordination could be extrapolated from the seemingly less emotional aspects of girls socialization process. However, in patrilocal societies, daughters are often temporary members of their natal group. The development of strong emotional and dependent ties with a daughter could increase the stress of her inevitable departure. A mother may, therefore, expend differential emotional investment in daughters and sons (see the following discussion).

At the same time, a daughter is not solely dependent upon her mother for social and domestic learning or for emotional support. A young girl may spend a considerable amount of time observing or being with her mother's social network (Wolf, 1974), from which she is likely to acquire information on child care, food producing and preparing techniques, husband-wife relationships, health-care, and a wide variety of other sociocultural, reproductive, and subsistence information. By contrast, sons are permanent residents of their household and parents are often partially dependent on their sons during their elder years. This makes it advantageous for a mother to nurture long-lasting bonds with male offspring, which is accomplished through intensive and emotional socialization processes (Wolf, 1972, 1974). Mothers may also invest differentially in sons because there is relatively little paternal investment in the early childhood socialization process: fathers spend a great deal of time in the rice fields, and there is a traditional tendency for "formal" father-son relationships (Wolf, 1974). The more demanding physical and emotional socialization process of girls relative to boys, plus patrilineal descent and Western ideals (that place a heightened value on economically based subsistence roles relative to child care and other domestic responsibilities) has contributed to the belief that male children are more valued than female children in Taiwan and many other patrilocal societies. However, this view does not consider all factors that reflect the value of a child.

Sex preferences for children should instead be a function of the number and sex ratio of existing offspring and the time lag between parental investments given to offspring and the offspring's reciprocation back to its parents. For example, a newlywed woman in Taiwan should desire a son because his birth will alleviate pressures from her in-laws to produce a male descendant who will assure their ancestral worship (Wolf, 1974; Johnson, 1975). A father, on the other hand, may be eager for a son who will eventually provide farming assistance, as well as inherit his land. These benefits from male offspring, however, will not be totally realized until he marries as an adult. When a grown son marries, his bride win be of great assistance to his mother.

Furthermore, adult sons typically provide their elderly parents with political, economic, and emotional security (Wolf, 1972). These factors predict that younger parents will tend to prefer sons over daughters. Conversely, the anticipated domestic contributions from a daughter suggest that a woman should have a strong desire for a female child after she has produced at least one healthy son. In other words, girls may be as valuable to their families as sons, but in different ways. Sons are more helpful in their later years, whereas daughters are more immediately beneficial to their parents during their earlier developmental years. Daughters contribute to their families in numerous ways. They perform many domestic tasks at a relatively young age, such as sewing, cooking, sib caregiving, and cleaning (Wolf, 1975). By performing these activities, daughters are contributing to the hygienic and nutritional health and cohesion of the, basic Taiwanese social unit, the family.

A daughter's contribution is further exemplified when mothers become ill, as the latter are usually unable to escape daily responsibilities even when sick (Kleinman, 1980). Since a woman's social value is often correlated with her ability to fulfill these responsibilities, thereby maintaining a cohesive family unit (Wolf, 1972; Ahern, 1978; Kleinman, 1980), failure to do so can be most stressful. In fact, women most often commit suicide for family-related reasons (Wolf, 1975), whereas men tend to do so for politically-related reasons (King, 1975). Thus, without supplemental help from a competent daughter, a mother's health and well-being could be further jeopardized during illness.

Due to the many contributions of a daughter, a woman's reaction to the birth of a girl is predicted to be a function of her age, as well as of the number and sex of her other children. However, these views have yet to be put to empirical test. The ways in which a daughter learns and uses her skills influences the health, cohesion, and stability of her uterine family. They also influence the quality of the husband to whom she can be matched and her progress into the next life stage. Presumably, similar evaluations are taking place with respect to the qualifications of her potential groom. When marriage takes place, the woman, moves in with her husband's family, a transition which embodies the true test of her competencies.

B. Stage Two The Wife

Moving into her husband's home may be stressful for a young woman. Unlike her husband, she is suddenly isolated from her own family and must somehow adjust to a new social environment. The adjustments will be especially difficult if conflicting interests develop between the bride and her new mother-in-law. These mother-in-law and daughter-in-law conflicts also appear to have contributed to the interpretation of female Subordination. Thus, it becomes important to consider what factors might account for such conflicts. Factors to be considered include (a) the mother-in-law's defense of her vested interests in her son; (b) the bride's vested interest in herself and her future; and, (c) the relative reproductive potential of the two woman.

The following discussion will address these issues, as well as the possible ways a new bride might ease her adjustment into a new living environment. When a young woman moves in with her husband's family, she benefits the mother-in-law with her domestic contributions. Such benefits may not, however, outweigh the threat to the mother-in-law of sharing her son and home with another adult woman (Wolf, 1975). The mother-in-law has invested a great deal in her son for future economic, social, and emotional security. Thus, the presence of a bride may be perceived as a psychological threat to the family structure and its function (Wolf, 1972). This initial phase of the patrilocal transition often creates a potentially stressful atmosphere for both the wife and the mother-in-law.

Over the long term, this stress could have an impact on the inclusive fitness (Hamilton, 1964) of both women; continual psychological (or physical) stress can disrupt a person's health, child-rearing abilities, personality, sexuality, and maintenance of interpersonal relationships (Kleinman, 1975). However, these two women need not, necessarily succumb to the consequential stress if a conflict does develop between them. For example, the mother in-law is likely to receive emotional support from other family members. Her husband, married son and any other family members in the home are likely to favour her in conflict situations because of their long-term kinship bonds and desire to protect the family structure and its function.

The concern for, and protection of, such family interests may have contributed to the social norms, for example, that forbid open displays of affection between wives and husbands (e.g., Irons, 1979). Alternatively, Hsu (1965) has suggested that restrictions on public behaviour between spouses were established for the maintenance of public female subordination. However, Hsu's view ignores the likely possibility that wives may also have vested interests and ways of protecting themselves from such stresses. For example, a wife can disrupt her husband's perception of family cohesion by trying to divide the loyalty between him and his mother (Wolf, 1972, Ahern, 1975). A woman can also disrupt her husbands sociopolitical authority through activities in her developing female social network (see Section III, C). A compounding consideration regarding the possible conflicts between a wife and her mother-in-law concerns their relative abilities and desires for future offspring. Since the wife win be bearing the mother-in-law's grandchildren, inclusive fitness theory predicts the existence of a reproductive conflict between them as long as the latter's reproductive value is more than one-half that of the daughter-in-laws.

The young bride's reproductive potential should be quite high at the time of her marriage. Thus, if the mother-in-laws reproductive value is still high and she still desires more young, the brides presence could facilitate reproductive competition between them. Inclusive fitness theory, fore, predicts that conflicts between a bride and her mother-in-law should correlate with the mother-in-laws age, and hence her reproductive value. Another consideration regarding reproductive competition between these two women includes the consideration of other adult sons in the home. If the mother-in-law has only one son and, therefore, only one daughter-in-law, the two women's overlapping interests could provide the impetus for a congenial relationship.

On the other hand, if there are other married sons, such conflicts may be predictably high. The overall potential for interpersonal conflicts in the family does seem to increase with the number of other daughters-in-law in the home (Wolf, 1972). At the same time, if all of the married sons and their wives have comparable reproductive potentials, the mother-in-law should have equal interests in all of the daughters-in-law and their children. The daughters-in-law may thus strive for harmonious relationships with one another to ensure minimal domestic conflicts. In fact, for this very reason, mothers-in-law should strive to reduce conflicts between their daughters-in-law. In spite of the potential for interpersonal conflicts, they do not always develop, and even if they do, they are likely to dissipate when mutually satisfying gains can be realized.

For example, a harmonious relationship between the bride and her mother-in-law could influence the relationship between the bride and her husband, thereby affecting their fertility potential. The couple's fertility potential affects the inclusive fitness of both women, making cooperation between them advantageous. A complementary or cooperative relationship between these women is also going to reflect the bride's acceptance into the local women's social community and in her overall credibility in the village. Since both of these women derive much of their emotional support and information base from their networks (Wolf, 1972) (see Section III, C), they both have much to gain by contributing to the network's cohesion.

The Relevance of Female Social Networks

Throughout much of Taiwan women frequently come together to perform both domestic and economically related chores (Wolf, 1972). During these gatherings, women typically exchange, transmit, and consciously use their individual and collective bodies of information for personal gain (Wolf, 1974). For example, women are most often responsible for family and personal health care. The responsibility is exemplified by the tendency for women more than men, to represent relatives at ritual healing sessions (Kleinman, 1980). Because of a woman's responsibility to maintain her family's physical and emotional harmony and because of the immediate consequences of her own ill-health, a woman should want access to the most available and personally relevant health-care treatment.

Presumably, the importance of the women's health-care role will encourage the exchange of personal or self-directed health-care techniques and information on traditional practitioners between women in social networks. These two forms of healing (i.e., personal or self-care and traditional practitioners and methods) are particularly important for women in their reproductive years (Ahern, 1975; Kleinman, 1980). Indigenous practitioner who emphasize public-style clinical, settings am most accessible to women, as well as to children and elderly populations both in rural and urban settings (Dunn, 1976). Traditional healing methods and practitioners also tend to link one's perceived symptoms with more culturally relevant causes (Pockert, 1976). This provides the patients with more meaningful diagnoses and treatments for their illnesses, as well as for continual self-care (Unschuld, 1976) and is therefore particularly relevant to those individuals responsible for their family's health care. Female social networks also allow women to affect local political, economic, and interpersonal matters, largely through their collective influences on male behaviours. Two of the primary ways of accomplishing this are through (a) the advice women' bestow upon their sons, thereby influencing the latter's decisions on investments and social affairs (Ahern, 1975) and (b) the culturally recognized phenomenon of "loosing face" (Wolf, 1974; Ahern, 1975). The second method is crucial to a woman's social influence because when a man loses face, his social, economic, and political credibility is devalued among other village members.

Losing face occurs through "gossip" between women in networks regarding a man's inappropriate behaviour. This "news" is eventually shared with the women's family members, which can be shameful to the man in question (Ahern, 1975). For example, if a man does not properly fulfill his role of father and husband, is abusive, or is in the process of making an unpopular political or economic decision, women can use their social networks as a threat to get him to either change or risk social alienation. In this way, women are not as isolated as a patrilocal system may initially project; they are also unlikely to be as vulnerable to physical and emotional abuse or total dependency on men as has been suggested in the past. In summary, female social groups in Taiwan are not meaningless (Wolf, 1972, 1974, 1975; Ahern, 1975). The information exchanged in such networks can contribute to family cohesion and health care and provides an avenue for female involvement in political, economic, and social matters. In these ways, female networks contribute to the survival and reproductive success of both family and other village members.

C. Stage Three The Female Elders

The third stage in a woman's life cycle is the "regal" grandmother role (Harrell, 1981b). In a general sense, post-reproductive women or grandmothers are honoured for their wisdom and are sought-after for advice on child care, health care (Alland, 1970), family problems, matchmaking, and religious activities (Wolf, 1974), as well as for money management and social affairs (Ahern, 1975). But many of these attributions went unnoticed until researchers recognized the value of women beyond their reproductive years (see also Harrell and Amoss, 1981). Assuming that a woman no longer has the physical capacity or emotional desire for more offspring, the presence of grandchildren will likely be a welcome change for her.

The freedom from obligatory or full-time offspring care, plus the domestic help from her daughters-in-law, will enable the older woman to focus more of her attention toward other matters, such as her social network activities and family relations. Such conditions allow a grandmother time to educate grandchildren on matters that are of particular concern to her. She can, for example, transmit her refined knowledge, to future generations, thereby contributing to socio-cultural transmission and development (cf Neisser, 1967).

An elder woman's social network is especially important at this time and is likely to increase the influence she has on her husband and other family and village members as well (Ahern, 1975). Relationships that women have nurtured with sons also become fully realized during this life stage. As adults, men more often seek and accept advice from their mothers on political, economic, and social issues (Wolf, 1974; Ahern, 1975). The opinions that men voice in the public arena then, are not necessarily just their own, but also represent those of their mother and her social network. In many ways then, women can be seen as subtly, but effectively influencing the socio-cultural developments of their family and perhaps even their entire village.

Changing Sex Roles, Health-Seeking Behaviours, and A Pluralistic Medical System

The chapter suggests that sex-specific strategies and roles have been quite complementary. However, the influx of Western, male-focused ideals and a changing economy in Taiwan have contributed to the devaluation of women's roles. Such male-focused value judgments will have unfortunate consequences for members of the less-valued sector of society. In Taiwan, this is particularly recognizable in the health-care system. Thus, the following section suggests that the devaluation of traditional health-care practices in Taiwan, which are now predominantly but not exclusively employed by women, is a consequence of a male-focused value system, primarily influenced by Western culture.

A. The Relevance of Traditional Medicine

Traditional Chinese medicine has been evolving for twenty-five centuries (Tseng, 1975). Since medical systems develop according to the ways in which a society is organized (Kunstadter, 1975) and in response to the perceived health needs of the people, traditional practitioners and patients should share similar medical concepts and beliefs. Although some traditional practitioners in Taiwan today provide advice on financial investments and subsistence activities (Tseng, 1975), their relevance is generally recognized in their ability to reinforce moral behaviours and return a patient and her or his family to health and order (Gallin, 1978).

Diagnostic sessions are common in which patients discuss problem and their sources in order to alleviate stress (Gallin, 1978) in ways that are culturally meaningful (e.g., public-style group sessions involving the participation of the patients, their relatives, and perhaps even a few neighbours). When an illness or stressful state originates from an interpersonal conflict with a relative or neighbour, these sessions can be revealing and even therapeutic. This is particularly important since culture influences our perception of what is stressful (Kleinman 1980), and stress can contribute to physical disabilities and even death (Sterling and Eyer, 1981). Cultural relevance in medical care can also be attained, in part, through doctor-patient agreement as to the causes of, and treatments for, an illness (Faberga, 1975; Zola, 1981).

In fact, a patient's decision whether or not to comply with a physician's treatment recommendation is often embedded in the ability of the doctor and patient to negotiate between their respective perceptions of the illness itself and its treatment (Katon and Kleinman, 1981). Such negotiations are best accomplished through insightful patient-oriented interviews a necessary skill that is accountable for up to 90% of accurate diagnoses (Twaddle, 1981). This requires that doctors use terms and concepts that are meaningful and familiar to the patients, which makes traditional Taiwanese healers particularly valuable to the more traditional members of society (Ahern, 1978). Belief in a cure and the ability to communicate shared beliefs In the cure have a strong effect on the success rate of the medical intervention (Katon and Kleinman, 1981).

B. Sex Differences in Health-Seeking Behaviours

Medical needs of people differ from one population to another (Harwood, 1981), according to both biological and sociocultural factors (Eisenberg and Kleinman, 1981). Individuals within a culture are also expected to differ in their health-seeking behaviours to the degrees that different amounts and types of medical information are available and acceptable to them (Chrisman, 1977; Kleinman, 1980). Westernization in Taiwan has brought with it Western medicine. At the same time, it has also allowed male roles to become westernized more quickly, giving men greater exposure to, and acceptance of this new form of medicine. The following summary describes how health seeking behaviours currently differ between women and men in Taiwan.

1. Female social groups appear to provide sources of traditional "female-specific health care information. There does not seem to be a parallel network system operating for men.
2. Women represent relatives at ritual healing sessions more than men (Kleinman, 1980), reflecting the greater role of the former in Taiwanese family health care.
3. Women are less able to escape daily responsibilities even when sick (Kleinman, 1980), and are, therefore, dependent on self-care, accessible practitioners, and reliable family members when ill.
4. Women represent the majority (70-80%) of temple visitors for aid from diviners (Tseng, 1975; Unschuld, 1976).
5. Women in general (Kleinman, 1980), but especially during their reproductive years (Ahern, 1975), are more likely to see traditional medical practitioners than men.

The preceding health-seeking behaviours of women reflect their vital role in maintaining familial cohesion and health care. However, older people of both sexes and the more traditional individuals, in general, are also inclined to seek traditional-style practitioners rather than Western-style practitioners (Ahern, 1978)

C. Reasons For And Consequences of Western Medical Domination

The use of traditional health care systems, combined with the increasing number of westernized biomedical practitioners in Taiwan, is contributing to its pluralistic medical system. Both systems are necessary as long as they serve the needs of the people. Nevertheless, the domination of the technologically advanced, but not always most appropriate, Western medicine is contributing to the reduction of available indigenous practitioners. If traditional medicine is still utilized by many individuals, then what accounts for the prevalence of Western medicine? Part of the answer lies in the identification of those most likely to use Western medicine; that is, men and individuals of higher socioeconomic status (Gallin, 1978). Since these individuals are also most likely to be elected as government officials (Gallin, 1978), their promotion of Western medical systems may represent attempts to fulfill their own health, economic, and political interests.

Furthermore, Taiwanese government support is almost entirely given to Western medical programs, and the provincial health department, which operates health clinics throughout Taiwan, is primarily staffed with individuals trained in Western medical procedures (Gale, 1978). These Western medical personnel and programs may be inappropriate for patients who adhere to traditional beliefs and values because Western medical concepts regarding illnesses and treatments often contradict those maintained by traditional patients (Ahern, 1978). These Western health- care orientations will inevitably be male-focused, as is the health-care system in the United States (Mendelson, 1981). There is little doubt that the emphasis on biomedical healing will have unfortunate consequences, especially for traditional women who retain their faith in indigenous health-care methods.

For example, some of the values traditionally attributed to grandmothers in Taiwan are jeopardized by the changing health-care system. Women elders have always been honoured for their indigenous medical knowledge. Thus, if the promotion of westernized medicine is reducing the credibility of those methods, female elders in Taiwan (much like the elders in many Western cultures) may soon feel the projected inadequacies of the knowledge for which they were historically honoured. Such changes may also have an impact on those who benefited from a grandmother's information to the degree that their knowledge also contributed to the prevention of, or treatment for common illnesses. A mother's crucial role in family-oriented and self-directed health-care delivery may also become devalued as a result of such westernization.

The consequential stress effects from its devaluation may then be further exaggerated by simultaneous loss of her options for traditional, stress-related treatments. The traditionally oriented women who prefer indigenous practitioners during their reproductive years may also face the possibility of reduced pre and postnatal care for themselves and their infants if traditional practitioners are unavailable. Of course, we must address the question of just how effective traditional healing methods are for present-day Taiwanese health-care problems before we can freely assess the consequences of its loss. The answer to that question can only be determined by careful comparative health-care studies.

One could also argue that individuals who choose traditional medicine only need to be educated or oriented to the concepts of Western medicine for it to become an effective method. However, the rising costs and urban-centered distribution of biomedicine reduce its availability for many potential patients. Furthermore, traditional socio-cultural (including medical) beliefs are often interwoven with child-rearing techniques, sex roles, religious concepts, and subsistence activities. To disrupt these belief systems further through a greater Western value orientation, may have consequences that lie far beyond the projected benefits of exclusively implementing biomedical systems.

6

SOCIAL BEHAVIOUR OF AQUATIC ANIMALS

HUMPBACK WHALES

At the turn of the century, humpback whales *Megaptera novaeangliae* became one of the main targets of the voracious international whaling industry. These playful whales were easy pickings for whaling ships. Humpbacks grow to 50 ft. long and weigh 40 tons or more. Fat content is lower than in other large whales but they are easier to find and catch. Their regular and easily located congregations for breeding in winter, and feeding in summer, meant that they were more exposed to hunting than the solitary blues and fins. In Antarctic waters alone an estimated population of 22,000 southern humpbacks was reduced to barely 3,000 individuals by the early 1960s when, thankfully, a ban on humpback whaling was introduced. The humpback whale, though, has a global distribution with migrations to the food rich Arctic or Antarctic waters in summer, and to the breeding grounds around the tropics in the winter.

In summer the northern race of humpbacks spend its time in the Arctic, filtering out small fish and shrimp-like crustaceans called krill. Apart from a few social interactions when meeting others, they are usually silent. It is after their migration south, to breed and calve in warmer waters, that their beautiful sounds are heard. In the Atlantic ocean, groups of whales can be spotted off the islands of the West Indies, around Bermuda and off the west coast of Africa. In the Pacific, southern California and Hawaii are known gathering places. This seasonal predictability and easy discovery nearly led to their extinction.

Since the turn of the century the worldwide population is estimated to have been reduced by 95% by whaling. Humpbacks are currently protected throughout their ranges by International Whaling Commission regulations. Except for limited aboriginal kills in the Caribbean and Green-land and inadvertent deaths when being caught in fishing nets off the east coast of North America, they are now safe. Ironically it was the whalers who first noticed that their prey was sensitive to sound. The disturbance of a noisy oar in its rowlock would frighten a whale away. There is even record of an underwater camera shutter release scaring a humpback whale 20 metres away. Their protection came not a moment too soon, for extinction would have robed us of a remarkable animal now known worldwide for its beautiful underwater songs. The songs were first recorded by O.W. Schreiber in 1952 from a US Navy underwater listening post on the submarine slope of Kauai, Hawaii, but he didn't recognise which marine creature was responsible. William Scheville at Woods Hole Oceanographic Institution later identified the sounds as coming from humpback whales.

The humpback subsequently became a world bestseller on an LP record called *Songs of Humpback Whale.* The scientist responsible for drawing scientific attention to their songs is Dr. Roger Payne of the New York Zoological Society who, together with his wife Katy, a musicologist turned zoologist, has been involved in a intensive study of humpbacks, their songs and behaviour. Roger Payne's interest in bioacoustics began with the directional sensitivity of bats' ears; continued with the owl's ability to locate prey by hearing; and then followed with studies of the way some moths can avoid bat sonar. His first encounter with whales was on a rain washed beach near Tufts University where a small porpoise had been stranded. Payne was appalled by what he saw. 'It had been mutilated. Someone had hacked off its flukes for a souvenir. Two other people had carved their initials deeply into its side, and someone else had stuck a cigar but in its blowhole. I removed the cigar and stood there for a long time with feelings I cannot describe.' He resolved at that moment to learn about whales, so that some day he might be able to have at least some effect on their future.

The opportunity came in 1967 when Roger and Katy Payne visited Bermuda to look at migrating whales and met Frank Watlington, an acoustics engineer at Columbia University Geophysical Field Station. Watlington was responsible for recording anything whether natural or man-made that produced sounds in the sea, in particular under-sea explosions. He played them underwater recordings of humpbacks and lent his tapes. While out in a row-boat, attempting to get close to humpback whales without frightening them, Roger Payne heard the sounds of whales amplified through the bottom of the boat. He noticed that one of the whales was singing phrases similar to those on Watlington's tape, and it suddenly became apparent to Payne that he was listening not to random sound but to regular repeating patterns. The humpbacks make a variety of grunts and squeaks too–indeed, they appear to have a whole repertoire of vocalisations–but these repeating patterns are by far the most interesting. They could be considered as true songs for they consist of long complicated repetitive sequences, much like bird song except that unlike most bird songs, each one can last from five to more than 30 minutes. (It even resembles bird song if speeded up.) A whale might stop singing or resume singing at any point in the song so it is difficult to determine a beginning, middle or end. Each song may be sung over and over again without breaks for many hours. A whale recorded in the Caribbean by Howard and Lois Winn, of the University of Rhode island, sang non-stop for 22 hours, breathing in the intervals between phrases.

It was still going strong when the Winns pulled up hydrophones and went home. Humpback whale songs are the longest and most complicated of animal songs known to man. Day and night the Paynes collected their sequences of whale songs by dangling hydrophones or underwater microphones from outriggers to each side of their small sailboat. The sounds were then recorded on tape-recorders and taken back to the laboratory and analysed with the help of Scott McVay at Princeton University. Together with Frank Watlington's recordings they have an almost continuous record of the Bermuda whales for a period of over 20 years. In an attempt to analyse the sounds, the Paynes divided each song into identifiable parts.

The smallest part is a unit (equivalent to a musical note). Units are grouped into small repeating sequences called phrases. Groups

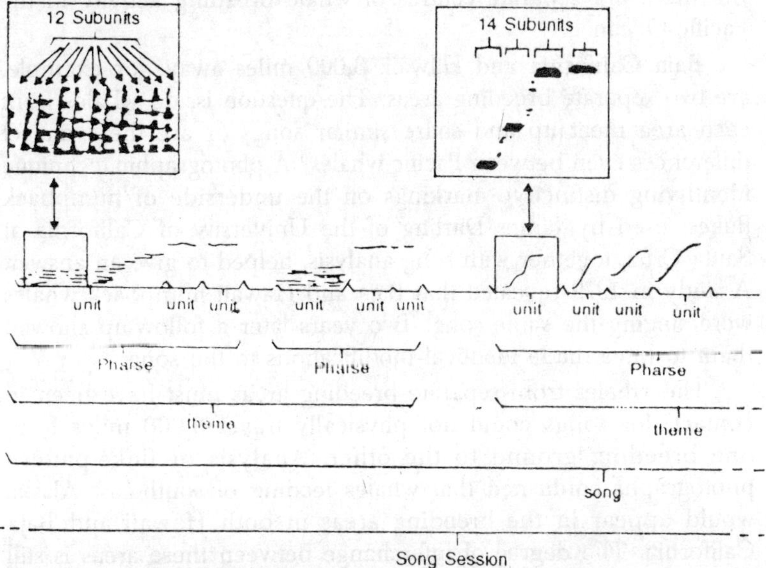

Fig. 6.1. Simplified humpback whale sonagram showing subdivisions of songs. (Insects show sections expanded if slowed down).

of similar phrases are known as themes. Off Bermuda, eight to ten themes make a song (later off Hawaii, as few as four to five themes were fond to make a song), and the song is repeated without pause as a song pattern. By breaking down the whale sounds in this way the Paynes were able to make a series of interesting observations. It turned out that all humpback whales in a particular area sing the same song.

Whales from the Pacific Ocean, however, sing a different song from those in the Atlantic, although the laws on which they base their songs appear to be the same in both populations. A song, for example, might consist of five themes (A-B-C-D-E) which always follow in the same order. If a theme, say C, is dropped, the remaining themes are always sung in the same sequence (A-B-D-E). In much the same way, the laws that govern the structure of sonnets, or the composition of western music are the same whether the piece is composed in New York or London. The laws governing humpback whale songs are, however, complicated and little understood and it is not known whether they are inherited culturally or genetically. The migration routes of the Atlantic and Pacific populations are unlikely to cross, hence the dialect- like differences.

But there are separate centres of whale breeding activity in the Pacific Ocean.

Baja California and Hawaii, 3,000 miles away, for example, are two separate breeding areas. The question is, do whales from each area meet up and share similar songs or are there dialect differences even between Pacific whales? A photographic technique identifying distinctive markings on the underside of humpback flukes, used by James Darling of the University of California at Santa Cruz, together with song analysis, helped to give an answer. A study in 1979 revealed that Baja and Hawaii humpback whales were singing the same song. Two years later a follow-up showed them to have made identical modifications to the song.

The whales from separate breeding areas must have been in contact, for songs could not physically travel 3,000 miles from one breeding ground to the other. Analysis of fluke-pattern photographs confirmed that whales feeding of south-east Alaska would appear in the breeding areas in both Hawaii and Baja California. The degree of interchange between these areas is still unknown. The discovery had other implications. It meant that whales might visit anyone of several breeding areas, a factor important for conservation. If, one year, a breeding area was affected by some man-made or natural catastrophe and all the whales killed, whales which that year had visited an alternative area might turn up after the site had recovered and repopulate the area. Also, when counting whales to establish stocks and whaling quotas, it would be quite possible to count the same whale twice–once in each breeding area–and get an overestimate of numbers. By studying whale sounds the researchers had identified a conservation loophole. Another surprising discovery was that the songs are continually changing.

All the whales in an open population change their songs in the same way so that each individual is up to date with the current vogue–a kind of top-of-the-pops. The changes are progressive and rapid, each component perhaps changing every two months, although many elements of the song are changing at any one time. Taking a folk-song analogy, in the first month an individual might sing: 'London's burning. London's burning. Fire-fire, fire-fire...' and so on, while two months later one phrase would be modified so the song becomes: 'London's burning, London's burning, London's burning, Fire-fire, fire-fire.....' etc. Humpback whales are composers, much as humans are, except that the whales do not create new songs, they evolve something different from what they already have by making minor modifications.

During a season a song may change components, add extra parts and drop in pitch. A component may undergo rapid changes for several months and then be left alone while other parts of the song are modified. Curiously, newly created phrases are sung more rapidly than older ones. Sometimes a new phrase is created by taking the first and last parts of older phrase and dropping the bit in the middle, much as we shorten 'I would' to 'I'd'. What selective advantage a whale may gain from changing its song is unclear. Whether it is dominant trend setter who introduces the changes will be difficult to find out, but what is clear is that an entire song is renewed totally after about eight years. Humpback whales sing only on the breeding grounds and occasionally during the migration. Only one isolated and misguided individual has been heard singing in the feeding areas in the Arctic or the Antarctic. What happens to the song, therefore, between breeding seasons? Whales, like elephants it seems, never forget.

In the new season the song is picked up usually as it was left the previous season, remembered by each male in the population. Most of the changes that take place occur during the singing season and not in between. (There is only one case of a substantial change in a theme, apparently between seasons). With such sophistication in song development, the question often asked is whether whales are highly intelligent creatures capable of indulging in intellectual pursuits and communicating in complex languages, even to the point, as some have suggested, of being wiser than man. True, whales have large brains, in fact the largest brains of any animals that have ever lived. How they use this large brain is unknown. It is tempting to suggest that whales reason about the world just as man does, but they are constrained by the limits of the environment in which they live. Whales lack hand and therefore brainpower cannot be channelled into the creation and use of tools, for example. They do, however, need to develop senses and skills to maintain their efficiency under the sea. Perhaps a part of their large brains is used in running this sophisticated sound system capable of discriminating songs above the backyard jungle of the noisy ocean, interpreting the content, learning the structure, remembering the detailed phrasing, processing the information, and may be modifying for future use.

The various frequencies of which a song is composed will attenuate differently in the water; higher frequencies will travel less far than low frequencies. Signals will sound quite different depending on whether the sender is near or far away. A whale would have immensely complicated computations to make in order to decipher the message as it was when sent. Clearly acoustic requirements in the sea could tie up a substantial part of the whale's brain capacity. (Bats, on the other hand, are involved in extremely complicated sound processing, using a brain of not more than one gram). At the very least, study of the humpback whale song is going to give some clues as to the way whale minds are working. Roger Payne's latest work is concerned with trying to relate songs to behaviour.

Already the whale researchers have noticed that, during certain phrases of a song, the humpback will make particular movements of a flipper or the body. Underwater photographer Al Giddings recorded seeing a singing humpback 60 fts. down moving its flippers back and forth in time with its song phrases. They seem to be dancing as they sing. Unfortunately, there are usually no other whales around to watch so perhaps they are merely flexing and stretching their muscles as they make the sounds. Whale song was originally thought to be associated with courtship behaviour at the breeding grounds. Writing in *Marine Bio-acoustics* in 1964 William Schevill noted, 'The sonorous moans and screams associated with migrations of (humpback whales) past Bermuda and Hawaii may be audible manifestations of more fundamental urges...' Whale renditions have been compared with those of birds, crickets and gibbons in that they probably convey such information as species, sex, age, location, identity, readiness to mate, and readiness to engage in aggressive behaviour with rivals. Aggression is not a word often associated with whales but it is becoming clearer that whales are more aggressive than was first thought. Researchers from the University of Hawaii have described 'raw freshly bruised dorsal fins' and 'head nodules that appeared red bruised'. Some have suggested that an assembly of singing humpbacks is reminiscent of a lek, where males gather on a communal courtship display ground to which females come in order to breed. It could be that females choose their mates on the basis of quality of singing, although there is no evidence of this as-yet.

Behaviour associated with singing whales has been studied by Peter Tyack of Rockefeller University. In 1977, while recording humpbacks around Hawaii with Roger Payne, Tyack would go out in a small boat, looking at random for whales, put the hydrophone over the side and record the sound when he could heard one whale dominating. 'When you get into the water near a singing humpback whale it can be a scary experience. Your lungs resonate with the sound; it is very loud and you feel it throughout your body–you feel it more than hear–it is very eerie to be sitting 30 ft below the water just having your body reverberate with the sound of a whale. Occasionally when one of us is alone in a boat and has been out there for a few hours, he'll notice the whale next to the boat, ten or 15 ft away. The whale will often surface, come right up within inches of the boat itself, seeming to look the boat over, often staying for up to half an hour at a time, spy hopping, lifting its head up to look into the look into the boat, lifting its flippers up; it is very strange to have your wild animal come up to you when you don't expect it'.

Often Tyack was able to identify the singing whale because, when it surfaced, the sound level would be reduced. Occasionally, he noticed, the whale would stop singing and then surface with another whale, the two whales moving off together. Following the whales in a small boat revealed little of their behaviour, so at the end of the season Tyack was left with little more than a tantalising hunch that something interesting was going on. In the intervening months before the next season, Tyack and his co-workers devised an observational technique which was going to reveal fascinating information about humpback whale behaviour, in particular some insights into the function of song.

The technique involved the combined operations of observers on a hill overlooking the whales, with others following individual whales in small boats (Boston whalers with outboards). A bay was chosen on the sheltered west coast of Maui, Hawaii, where large concentrations of whales appear each winter. In that way the overall patterns of movements could be seen from the hill while individual identities and vocalisations could be recorded in the boats. Each team was in radio contact with the other. In the spring of 1979 they tried their experiment. The first singer they followed stopped singing, joined with another whale and the two went off. During the season, 13 out of 28 whales that were followed after they had stopped singing joined with other whales. Indeed, the recording team learned to expect interesting things when a singing whale they were pursuing stopped singing. By observing whale behaviour in this way they were able to put together a picture of leviathan life below the sea. A singing whale is almost always alone, separated by several hundred metres from any other whales. It moves slowly, turning this way and that while singing.

It occasionally moves towards other whales but avoids any other whale that is singing. Singers do not appear to hold duets or to interact vocally in the way. They also do not seem to have strict territories. Singers are not found singing at the same 'song posts' on different days. In a few cases, Tyack observed singing whales approached by single non-singing whales. The original singer would stop singing, move along for a while with the intruder and then silently, but rapidly, swim away, seemingly displaced from his singing post. The new arrival would then start singing in the other's place. If a singer approaches a small group of whales such as cow and calf, he will often turn deliberately towards them. The cow and calf sometimes steer away, other wise they continue on the same course, but they rarely move towards the approaching singer. With a two year breeding cycle, where a cow may conceive one year, have a calf the next, and then wait for a year until the calf is weaned before ovulating again, a cow may not be sexually receptive, and so will actively avoid a singing male. The singer often pursues the group and will attempt to catch up. If he is able to join them he immediately stops singing and becomes, an escort to the cow.

The cow, the calf and the ex-singer then swim along slowly and silently together, the calf keeping close to the cow. Escort whales were once known as 'aunties' and were thought to be female helpers. Sometimes they will interact with gentle flippering or rolling, behaviour previously observed in association with sexual activity in gray and right whales. Normally any signs of aggressive behaviour are absent in cow-calf-and-escort groups. On one occasion, however, when Tyack was in the water with the whales, he somehow got between a cow and her calf, the calf having swum over to investigate Tyack. The mother, clearly upset, turned sharply towards Tyack and in so doing bumped into the surfacing escort whale which could not get out of the way in time. The escort produced a sound and the cow responded by emitting a series of grunt s for about half a minute which might have been translated as 'watch out where you're going!" Occasionally the cow and escort will dive below out of sight leaving the calf at the surface for a period of ten to 15 minutes, returning to the calf for five minutes and disappearing again. Whether mating takes place deep down in the ocean is not known, although researchers believe that this must be the case. Females ovulate at this time and males show an increase in testes weight (compared with the summer feeding condition) and increased sperm production.

In addition females calve in winter and the gestation period is about one year. Songs sung towards the end of the season tend to be longer than those at the beginning. A similar observation in songbirds has been accounted for by an increasing concentration of testosterone in the blood. It is thought that similar events could be happening with humpbacks. If the cow-calf-and-escort trio swim into the vicinity of another singing whale, the singer may begin to approach the group. Invariably the cow-calf-and-escort speed up and alter course away from the singer, but the singer seems highly motivated to join them. He will speed up dramatically, even while singing, and pursue the group, turning this way or that in an attempt to catch up. As soon as he has reached the cow-calf-and-escort he will also stop singing and a sudden change in behaviour takes place in the group. They begin to swim faster. The two escort

Fig. 6.2. Humpback whale cow-calf-escort group.

whales engage in some kind of competition in which the secondary escort attempts to displace the principal to escort from his position next to the cow. With the cow and calf up front and the two escorts behind the group may be swimming at ten to 15 km per hour. Until now the original trio were silent but as soon as aggressive activity starts, a whole barrage of sounds can be heard. The whales will thrust at each other with their flukes or ram into each other as they jockey for position. The impact of fluke on blubber can be heard as a very loud slapping sound.

Occasionally they blow bubbles underwater, an activity associated with aggression in many other species of whales. They also produce vocal sounds which the researchers recognise as social calls. There are many types of sounds–sometimes segments of songs out of context, high pure trumpeting calls and low grunts but as yet it is difficult to sort out their meanings. The social sounds are loud and can be heard up to nine km away, thus advertising the location of the group. Other lone singing whales, on hearing the disturbance, will stream into join the group. As more and more whales join, the activity increases. The rapidly enlarging group becomes very rowdy and engages in a great deal of aggressive behaviour. At the surface, flippers, flukes and heads can be seen being thrown out of the water.

Below, the group appears to keep a particular structure. The cow and calf are the centre of activity with the cow as the nuclear animal and the principal escort alongside. The rest of the group take up stations around the pair and continually challenge the position of the principal escort by attempting to place themselves between him and the cow, the 'nuclear animal'. By now there might be up to 15 animals in the group. Sometimes a secondary escort will displace the principal escort which will swim off. He may try to regain his position and is sometimes successful. However, the position of principal escort in a group is not long lasting.

According to whale researcher Hal Whitehead of the University of Cambridge, who first described the 'nuclear animal-principal escort' structure, the principal escort retains his position for only seven or eight hours. If one of the large groups has been swimming along together for a while, the activity seems to flag. Animals gradually lose interest and drop away. They take up well-spaced positions in the ocean and recommence singing. The cow-calf-and-escort group swims on. In order to confirm some of these

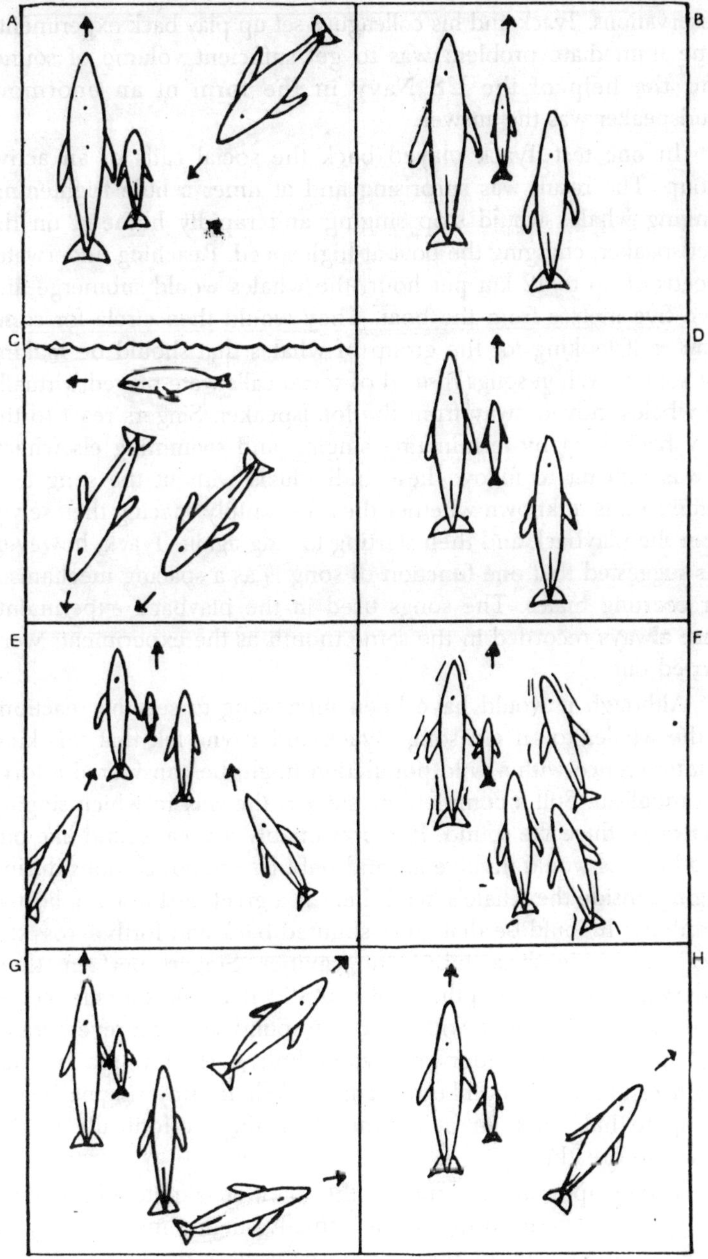

Fig. 6.3. (opposite) Humpback whale: cow-calf-escort groups.

observations. Tyack and his colleagues set up play-back experiments. One immediate problem was to get sufficient volume of sound, and the help of the US Navy in the form of an enormous loudspeaker was the answer.

In one test Tyack played back the social calls of an active group. The result was surprising and at times a little frightening. Singing whales would stop singing and rapidly home-in on the loudspeaker, charging the boat at high speed. Reaching underwater speeds of up to 12 km per hour, the whales would submerge and dive five metres from the boat. They would then circle for some time as if looking for the group of whales that should be making the sounds. When songs instead of social calls were played, virtually all whales moved away from the loudspeaker. Singers react to the play back song by terminating singing and swimming elsewhere. It was difficult to follow these individuals without the song as a guide so it is unknown whether they are simply spacing themselves from the playback and then starting to sing again. Tyack, however, has suggested that one function of song is as a spacing mechanism for courting males. The songs used in the playback experiments were always recorded in the same month as the experiments were carried out.

Although it would have been interesting to see the reactions of the whales to an old song. Tyack and Payne felt that this kind of interference with a wild population might be considered a form of vandalism. Still a complete mystery is the way in which singing whales produce the sound. It is presumably a vocal sound like our own but this would involve air and bubbles are not obvious during singing. Inside the whale's head there is a great deal of complicated plumbing. It could be that air is shunted back and forth across the vocal chords in these tubes and cavities. Singers perform their underwater arias at depths of 80 to 100 ft below the surface. A deep submarine canyon will reflect the sound so that it reverberates as if in a gigantic underwater cathedral. A shallow sea bottom produces almost studio-like recordings. Whales stay singing below for up to half an hour at a time, returning periodically to the surface to breathe.

As they approach the surface the sound gets quite a bit fainter. Whether this is due to the whale actually singing more quietly or caused by some acoustic property of sea water is not known. When breaking the surface they do not stop singing or break the rhythm of the song, but during four or five identifiable pauses in the song they will take breaths before diving below once more. There is a particular song theme in which breathing tends to occur so Tyack and his colleagues were able to anticipate a singer's arrival at the surface. As they tip their flukes and sound, the volume of the song becomes much louder, literally vibrating through the boat. In the feeding grounds in the polar region the whales are relatively quite. They do, however, occasionally indulge in social conversations using sounds similar to those emitted by the large rowdy groups.

During a preliminary study in the Glacier Bay area of Alaska, Tyack observed that the sounds are usually made when the whales are in groups, particularly when individuals meet and join or split up. Bill Dolphin, at Boston University, has made a more extensive study but is not sure whether the sounds are greetings or keep away calls. Whales frequently feed in organised groups, sometimes in line or chevron formations, so competition for resources is unlikely to be the reason for any spacing calls. Social calls could be used to keep a feeding group together but there are no observations of this as yet. Humpbacks have one usual form of feeding in which sounds are often heard, called bubble-net feeding. Starting low in the water, the whale will lay a circular column or net of bubbles in the water.

As the circle is closed the whale will come up right in the middle of the bubble tube engulf the food in the centre. Tests on the density of food in the bubble net have shown it to be up to 50 times higher inside the circle than in the surrounding water. Pleats in the lower jaw swell out with each mouthful. The whale's jaw is then raised and the pleats flatten forcing the water out between the baleen plates and leaving a concentrate of krill and small fish which can be swallowed. According to Charles and Virginia Jurasz from Glacier Bay, Alaska, the diameter of the bubble is selected and nets with a variety of mesh sizes can be made.

Southern Right Whales

While Peter Tyack and his colleagues studied humpbacks off Hawaii, another group of Rockefeller University whale-watchers, working with Roger Payne, directed their attentions to another great gathering of whales, this time southern right whales *Eubalaena australis*. Reaching lengths of 50-60 ft these enormous blue-blacked whales are characterized by the 'bonnet', a collection of white, crusty patches on the top of the head and on the snout which often provide anchorage for whale lice, barnacles and sea anemones. These callosities, as they are known, probably serve similar functions to human eye-brows and other facial hair, in the case of the whale deflecting water from entering the blow-hole when at the surface. Those on the snout seem to function as 'antlers' during aggressive bouts. The centre of activity was the Peninsula Valdes in Patagonia–an enormous cape enclosing two shallow, almost land-locked and very desirable bays; desirable for whales that is, for Peninsula Valdes is renowned for its high winds and rough seas. The windier it is better southern right whales like it.

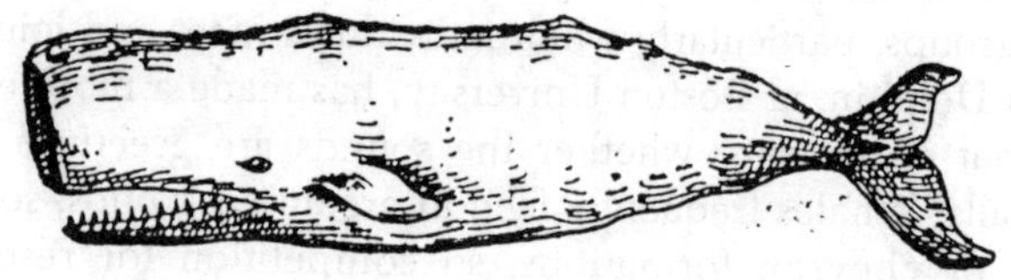

Fig. 6.4. Physeter (Sperm whale).

As the wind increases and white-horses appear on the sea's surface, the whales burst into life and literally play in the storm. An individual will leap from the water, crashing back in an explosion of spray. Sometimes a huge tail is seen sticking vertically out of the water. Observations have revealed that southern right whales actually sail in the wind. High winds are associated with much activity including lob-tailing (slapping the water with the tail fluke), flipper-slapping and breaching. During a storm, underwater noise increases, particularly as a result of waves crashing on the shore, thus making the lower frequencies with which southern right whales communicate. By slapping the surface, individuals can keep in touch. Investigating sound communication and its importance in the social behaviour of southern right whales are Rockefeller University's Christopher and Jane Clark.

Originally a biochemical engineer, Chris Clark got involved by accident–he let Roger Payne borrow his pick-up truck, and after a few visits to the Payne household became hooked on whales. Roger Payne and his family had already spent five seasons recording southern right whales at Peninsula Valdes, and their enthusiasm was clearly infectious. Chris Clark's expertise in electrical engineering was to help solve the problem of following whales under water. Using an underwater hydrophone array linked to a portable mini-computer, Clark and his colleagues were able to observe the whale's movements and interactions from the cliff-top while at the same time plotting the positions of the whales making sounds. The most prolific sound to be heard in the Bay is a simple low frequency, tonal call which is thought to be used for contact over long distances. As individuals approach one another they exchange an increasingly rapid series of calls, and having met stop calling altogether. These calls can be heard at anytime of the day during the five months of the year in which the whales congregate in the bays.

Contact calls have a frequency between 100 and 200 Hz, which happens to coincide with the quietest range of frequencies in the usually noisy bay. Southern right whales in the bay appear to be using a low-noise sound window through which to communicate and so make contact with others over great distances with the minimum of effort. Whales calling in the bay, a shallow cathedral-like dish 25 kilometers long by 15 kilometers wide, can be heard plainly from one side to the other. In deeper water the sounds should travel even farther. Often, several whales will come together to form a tight active group. Five 40-ton males and three females, for example might be swimming in an area the size of a gymnasium. In these active groups the sounds tend to become more excited, rising in pitch, with mixtures of high melodic notes and low, pulsed growls–the higher the pitch, the higher the excitement.

It is difficult underwater to work out which whale is making which noise, but it is already clear that when a large number of males get together aggressive growling sounds dominate proceedings. It has been suggested that they are competing for the female, prior to mating. If two males are left with a female, bouts of growling are heard. Eventually one male departs and the sounds change to more high-pitched melodic notes. Chris Clark describes the vocabulary as a continuum, rather than a set of discrete sounds except, that is, for the stereotyped contact call. Curiously southern right whales don't have an alarm calls, although frequently harassed by killer whales. Indeed, the calls seem not to contain any complicated messages. Right whales are grazers, and therefore do not require a sophisticated communication system to coordinate activities such as hunting. They are promiscuous rather than monogamous, so few social sounds are required. Their only need for calls is to bring the 'herd' together. Once he had identified the southern right whale's repertoire and suggested possible functions for the sounds.

Chris Clark used play back experiments to determine whether the interpretations put on the calls were indeed correct. His first task was to see if whales responded to contact calls. A variety of sounds was collected for the experiment. In addition to the right whale contact calls themselves, Clark tried background noise, pure tones at frequencies contained in whale calls (i.e. 200 Hz), humpback whale songs, and even Handel's Water Music! An underwater loudspeaker was placed on the bottom of the bay, not far away from the hydrophone array, in front of the observation hut. To make identification easier, the observers waited until there were only two or three whales in the area. A whale would be spotted by the pattern of callosities on its head and its calls recorded for a quarter-of-an-hour before the playback experiment. Playbacks were started when the whale had passed the loudspeaker and hydrophone array, and was heading out of the sea. The observers noted any sounds the whale made in response to the playback, and its direction and speed. The whales responded only the playbacks of southern right whale calls. They would stop, turn around dramatically in a flurry of foam, and swim at full speed back to the loudspeaker, at the same time producing more sounds. When other playbacks were used, whales would ignore them and continue to swim away.

On one occasion, at the end of the season, when few whales were in the bay, a lone right whale was exposed to playback of a group of whales. It responded by returning to the speaker. No more sounds were played and the animal began to swim off. After a little while in turned back and swam to the area of the speaker. It continued to do this until dark when Clark left it alone in the bay silently searching for its fellows. Having shown that southern right whales can differentiate their own calls from other similar sounds, the Clarks hope to continue playback experiments in order to determine the biological functions of the sounds in a whale's acoustic repertoire.

Scattered Herds

For many years a mysterious low frequency sound has been heard in the oceans of the world. The sound is pulsed in 'blips' with a frequency close to 20 Hz. With a bandwidth of only 3 Hz, the '20 Hz signal', as it is known, is almost pure tone. It is also of

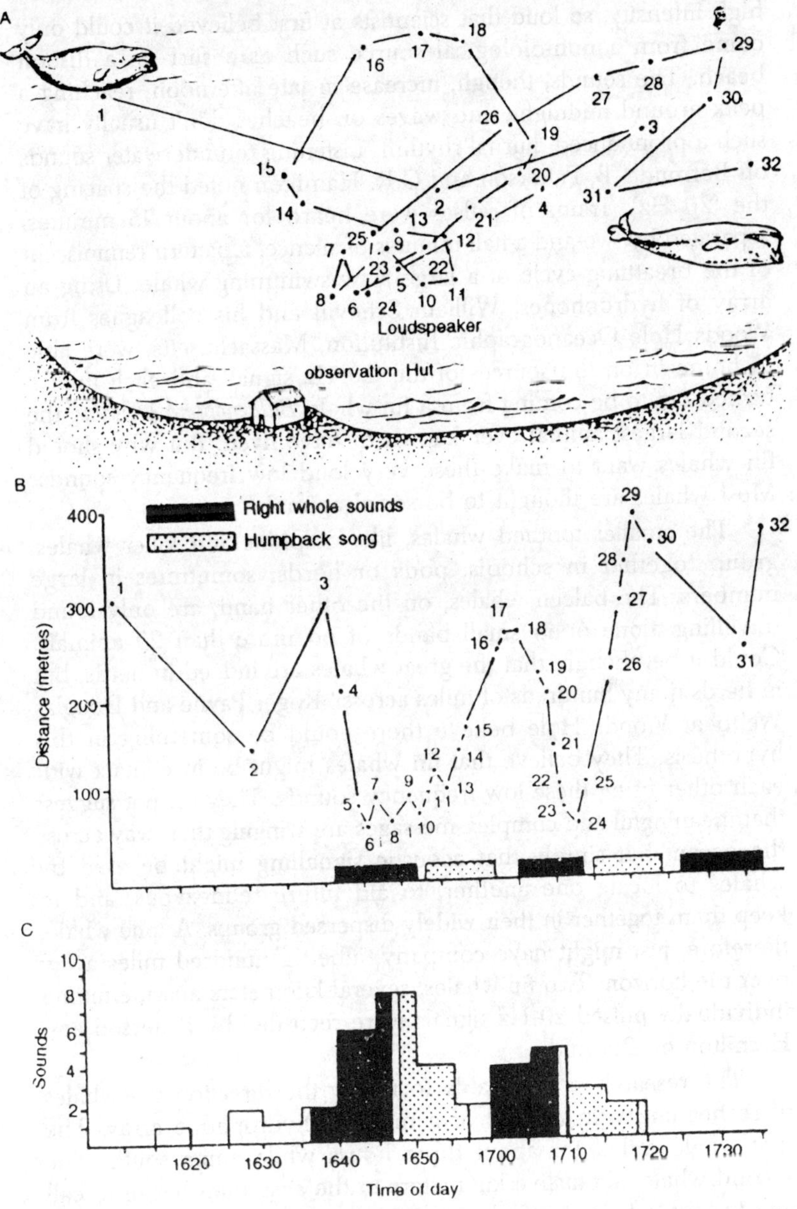

Fig. 6.5. Response of Southern right whale to the playback of right and humpback whale sounds: A–The path of the whale. B–The distance of the whale from the loudspeaker. C–The number of sounds made by the whale per 5-minute interval.

high intensity, so loud that scientists at first believed it could only come from a non-biological source such as a surf on a distant beach. The sounds, though, increase in late afternoon, reaching a peak around midnight, and waves on beaches don't usually have such a pronounced diurnal rhythm. Listening to underwater sounds off Bermuda, B, Patterson and G.R. Hamilton noted the spacing of the '20 Hz'. Trains of pulses were heard for about 15 minutes, separated by two-and-a-half minutes of silence, a pattern reminiscent of the breathing cycle of a large slow swimming whale. Using an array of hydrophones, William Schevill and his colleagues from Woods Hole Oceanographic Institution, Massachusetts, were able to home in on the sources of the '20 Hz signal' and each time it turned out to be coming from a fin whale *Balaenoptera physalus*, the second largest animal ever known to have lived. But why should fin whales want to make these very loud low frequency sounds? Most whales are thought to be social animals.

The smaller toothed whales, like dolphins and killer whales, group together in schools, pods or herds, sometimes in large numbers. The baleen whales, on the other hand, are only found travelling alone or in small bands of no more than 20 animals. Could it be, though, that the great whales are indeed in herds, but in herds many hundreds of miles across? Roger Payne and Douglas Webb at Woods Hole believe there could be something in this hypothesis. They believe that fin whales might be in contact with each other using these low frequency sounds. They do not suggest that meaningful and complex messages are winging their way across the ocean, but simply that acoustic signalling might be used by whales to locate one another, to aid future rendezvous, and to keep them together in their widely dispersed groups. A lone whale, therefore, just might have company, albeit a hundred miles away over the horizon. Two fin whales, several kilometers apart, emitting individually pulsed 20 Hz signals were recorded by Patterson and Hamilton off Bermuda.

The researchers were able to follow the direction the whales were heading with the aid of a multiple hydrophone array. The first whale called for about three hours while going south. The second whale, about five kilometers to the east, then began to call and the first whale changed direction towards it. Was the first whale being guided to the second by sound? Follow-up work in this area is prohibitively expensive so as yet there is no answer, but the

Fig. 6.6. Blue whale and Harbor porpoise.

simple and repeated patterns of pulsed sounds are ideal for long-range communication.

Fin whales, unlike humpbacks and grays, do not appear to breed always in the same areas, so there must be a mechanism to bring individuals together. The 20 Hz signal seems to fit the bill. It is lower in frequency than the noise generated during turbulent storms; loses little energy when bounced off the sea bottom; and is apparently the best frequency for propagating through a surface-frozen polar sea. But the ocean is a noisy place. The whales would have to make themselves heard over considerable background chatter, particularly in today's ocean where man's supertankers and other technological advances pollute the ocean not only with oil but also with noise. How do they do it? There is one published case of a small underwater explosion from just four pounds of dynamite, detonated off Australia and detected at Bermuda at about 12,000 miles away. That is unusual, but calculations made by Payne and Webb indicate that fin whales could communicate over considerable distances by making use of deep water sound channels. These acoustic channels are the result of physical characteristics of the ocean. Differen .es in water densities, salinity, temperature and ocean currents, produce a channel at a particular depth which tends to trap sound. It works as a kind of underwater voice-tubed by concentrating sound energy in a narrow beam rather than diffusing it over a large area. Without the help of a deep ocean sound channel, individuals could speak with one another over a distance of about 50 miles. Making use of a cylindrical propagation in a deep water sound channel, the distances could reach a maximum of about 500 miles.

In pre-steamship days, these distances would have been further increased to 140 miles and 3,500 miles respectively, and these are conservative estimates. Clearly the fin whale acoustic system must have evolved in quieter ocean and it is conceivable that man's maritime developments could have seriously impaired communication between whales and upset their lifestyle–a further set back following their wholesale slaughter at the beginning of this century. So far, researchers have not demonstrated that whales actively seek out the sound channel, but inevitably their calls will spill into the channel and the sounds propagate over long distances, audible to any whales listening at channel depths. The sensitivity of hearing in baleen whales is thought to be excellent albeit in the lower frequencies. Fin and humpback whales have many more fibres in the nerves from the ear to the brain either man or the bottlenose dolphin. They appear to lack the high-frequency or ultrasonic hearing capabilities of dolphins but they may hear very low sounds in the infrasonic frequencies.

The blue whale *Balaenoptera musculus*, for example, the largest creature ever known to lived on earth, is thought to emit very low frequency grunts, but surprisingly there has been suggestion that it is also capable of vocalising in ultrasonic frequencies. The minke whale *B. acutorostratus* one of the smallest of the baleen whales, calls with a very low note. The Californian gray whale *Eschrictius robustus* makes moans, bubble sounds, knocks and 'a metallic - sounding pulsed signal'. They also produce clicks which are probably used for navigating and locating food. The sperm whale *Physeter catodon*, a large odontocete or toothed whale, produces clicks which are thought to travel long distances under the sea. William Watkins and William Schevill at Woods Hole have been analysing the sounds made by sperm whales. They have a multiple array of hydrophones and can locate accurately the positions of individuals.

Groups of these large, toothed whales are heard to make clicking sounds and each whale produces its own distinctive pattern of clicks like a morse code.

Using this unique identification and location technique, Watkins and Schevill have been able to track individual sperm whales and observe their behaviour. They have found, for example, that sperm whales surfacing within ten meters or so of each other spread out like an inverted funnel when they dive again, and so are separated by much greater distances as they reach the bottom. Returning to the surface they emit more clicks and gather together once again in a close-knit group. Sperm whales are more obviously gregarious than some of the other large whales, and their click signatures would be a important in a complex social structure where individuals might want to contact each other during mating, feeding and so on. Groups of sperm whales are thought to be led by a dominant animal which would use sound to guide its family group.

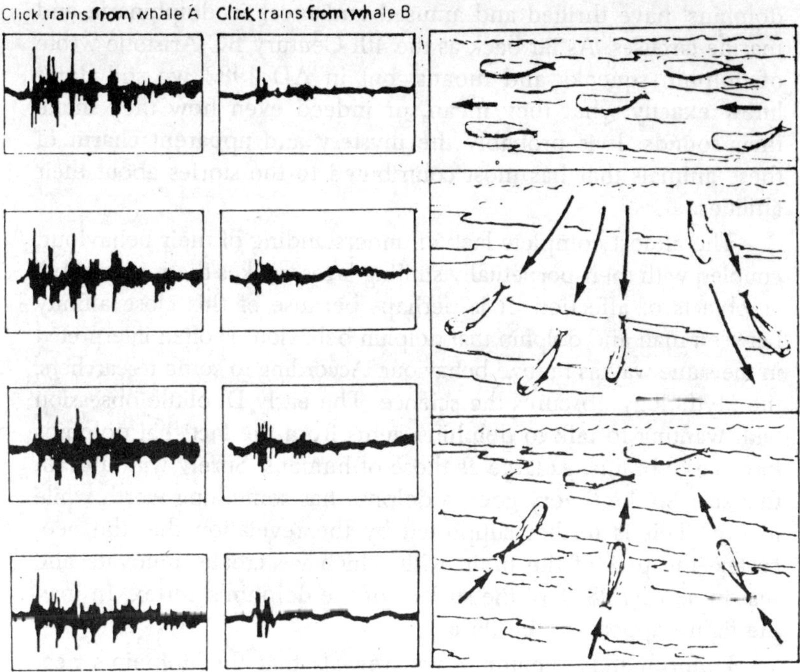

Fig. 6.7. Sperm whale herd on the move, each individual producing its own sound signature of clicks.

Whalers were aware of messages passing between sperm whales. They were sure that distress or alarm calls were given by harpooned animals which seemed to elicit responses from others many kilometres away. Although we still know surprisingly little about the behaviour of humpback, southern right, sperm, fin, blue and minke whales, we do know that, despite international agreements, they are still threatened by the activities of our own species. By the time we understand sufficiently about them to begin the urgent work of conservation, it may be too late. The damage may be irreparable. In the future their beautiful calls might exist only on long-playing records and the research tapes of scientists.

Dolphin Language

Killer whales, pilot and sperm whales, dolphins and porpoises are all classified together as the odontocetes–the toothed whales. Characteristically they are all noisy animals, but the voice of the dolphin is perhaps the most familiar. All over the world, 'singing dolphins' have thrilled and amused audiences in dolphinaria and marine circuses. As far back as the 4th Century BC Aristotle wrote of dolphin squeaks and moans, but in AD 1983 we still don't know exactly what they mean, or indeed even how they make their sounds. It is probably the mystery and apparent charm of these animals that has most contributed to the stories about their abilities.

The almost complete lack of understanding of their behaviour, coupled with their perpetually smiling faces, makes them irresistible as objects of affection. It is perhaps because of this close affinity between man and dolphin that dolphin behaviour is often interpreted in the same way as human behaviour. According to some researchers, the mythology obscures the science. The early Doolittle obsession with wanting to talk to dolphins stems from the fact that dolphins have large brains, as large as those of humans. Surely with a brain that size, so the theory goes, a dolphin has something worth while to say? This is further supported by the revelation that the neo-cortex–the part of the brain with which we create, innovate and reason–covers 98 % of the surface of the dolphin's cortex. In man the figure appears to be 96%.

Unfortunately, anatomy also reveals that the dolphin's neo-cortex is much thinner than in humans, so this line of reasoning is rather inconclusive. It was thought at one time that, contained within the dolphin's wide range of vocalisations, were complicated messages that could rival the complexity of human speech. Dolphins were thought to have mystical, paranormal powers and were attributed with super intelligence. They were thought to be among the cleverest animals on this planet. Doyen of the 'smart dolphin movement' is John Villy, a doctor who has studied such topics as neurophysiology and hallucinogenic drugs. His object has been to discover a means of communication between dolphin and man in order to understand dolphin language, culture, philosophy, and even their system of ethics. In an early experiment, Lily had dolphins mimic English. They could follow a count up to ten, and almost say some simple English words.

The recordings are amusing, but mynah birds and budgerigars can do better. Lilly's early mimicry experiments, however, did turn up some interesting information. Dolphins, it seems, showed a remarkable ability for rapid mimicry on some occasions an animal would begin mimicking before the signal it was copying was finished. The time domain, therefore, was quite different from that in budgies. In addition, in another experiment a researcher would read out a series of syllables, consisting of vowels and consonant sounds, and the dolphin would mimic the same number of sounds with the same durations with 70% accuracy.

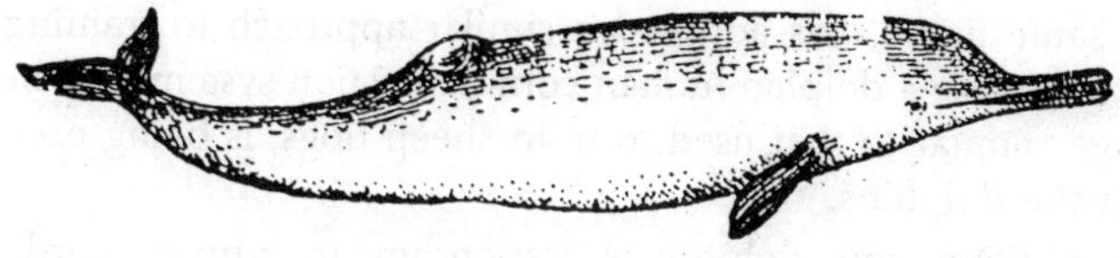

Fig. 6.8. Platanista gangetica (Dolphin).

Dolphins obviously appreciate quantitative aspects of human speech much better than do mynahs and parrots. English, though, was clearly not the language with which to investigate the linguistic abilities of these remarkable creatures. Today, teams from California, Florida, Hawaii, and The Netherlands are carrying out numerous costly experiments using signs, symbols and computer-assisted sound equipment, all witness to man's need to communicate with another animal. There is, of course, a serious side of this kind of research. By 'conversing' with the study animal, information can be sought about its cognitive characteristics and an understanding gained, perhaps, about its intellectual abilities and limitations. Lilly himself is involved with the JANUS (Joint Analog Numeric Understanding System) project in California.

A sophisticated computer-assisted sound system receives and transmits bleeps and whistles to and fro between captive dolphins and their human researchers. Dolphin sounds are matched to computer generated sounds, and to visual letters and other symbols which the dolphins can see on an underwater screen at side of their tank. The dolphins react to the symbols on the screen, cause the sounds or symbols to change or simply match the sounds with dolphin sounds. The aim of the experiment is to create a new form of 'language' that is mutually accessible to both man and dolphin. Tests have started with 48 or so sound and symbol combinations, known as morphemes. Each sound symbol combination is associated with an object, a place or an action. Lilly is using the computer interface to link the dolphin's high frequency (1,400-4,000 Hz) vocalizations with the human's relatively low frequency (200-2,200 Hz) speech and to change the time domain so that the dolphin's utterances can be directly translated into the relatively slow delivery of human speech. Lilly estimates it will take about five years to work out a human-dolphin dictionary which could be used to communicate across the species boundary. Flipper Sea School in Florida., once host to the famous television dolphins of the same name, has adopted a similar approach to training in order to explore a dolphin-human communication system. A whistle language, similar to that used to train sheep dogs, is being used to instruct the dolphins.

In addition, one dolphin is responding to human words as well. At the Dolphinarium, Hardervijk, The Netherlands, Wilhelm dodoc Van Heel has been working with a Killer Whale called Gurden. Frequency modulated tones associated with objects in the tank were–played to the killer whale. When the signal was given Gurden was expected to touch the relevant object. This she did. If the wrong object was presented with a particular sound signal the whale became very upset. Eventually Gurden was able to imitate the object-linked signals and touch the correct objects. Van Heel them introduced sound signals, to represent action words, verbs, and the whale went on to recover objects on sound cues.

The remarkable thing about this particular whale, though, according to the television programme *Talking Whales*, was that Gurden began to talk back. During a training session she was heard to give the sound signal to fetch the object in the tank, a copy of the signal the researchers had given her. This was the first time the conversation had been two-way. In Hawaii, Dr. Louis Herman, director of the Marine Mammal Research Laboratory of the University of Hawaii, together with Dr. Douglas Richards and Dr. James Wolz, has been working on another major project with two female bottlenose dolphins *Tursiops truncatus.* The dolphins have learned to respond to visual and verbal signals in much the same way as a sheep dog; but there the similarity ends, for the dolphins talk back. The back ground to Louis Herman's current research interests includes many years of observation on the dolphin's sensory systems, particularly its capacity to see and to hear, and studies of its learning and memory abilities.

The dolphin, for example, has a good auditory memory, being able to remember long lists of sounds presented to it. Herman therefore wanted to know if language ('the most complex method known of information transfer from one creature to another') could be added to the growing list of dolphins' abilities. Instead of looking at the way a language might be produced by a dolphin, the research team concentrated on how an animal might understand language. To do this, Herman copied the technique used by second-language teachers who instruct pupils by getting them to carry out tasks. The level of understanding can be measured by how well the task is carried out. This involved entirely a new method of training. At a dolphinarium a normal way to get an animal to learn a new tricks to entice it gradually with food. To teach a dolphin to go through a hoop, for example, the animal is presented with a hoop, rewarded initially for poking its head through, and then rewarded progressively as more and more of its body passes through until it swims through completely.

A gesture or sound signal would then be provided to initiate the entire behaviour of swimming through the hoop. Herman's technique differs in that the dolphin is first taught the word for 'hoop' and then the word for 'through'. The words can be combined into 'through hoop' and without special training the dolphin will know what to do. If a new word such as 'gate' is introduced and linked to 'through' the dolphin knows to go through the gate as it did the hoop. An object word and an action word are taught to the dolphin and combined with other words in new contexts; this gives the researchers flexibility in the training system. Unlike sign-language experiments with primates, where chimps and gorillas are invited to converse with the trainer, thereby producing 'words' which might be open to misinterpretation.

Herman is simply examining the cognitive characteristics of dolphins with what he feels are more objective interpretations of the results. Herman is simply asking questions. In what do dolphins specialise? What are their limitations? What are they good at? How do their success and failures compare with those of other animal? What role does intellect play in a dolphin's world? In short, Herman's main object appears not to be language but to create a tool for discovering a whole range of behavioural abilities in dolphins.

Dolphin Sounds

What of dolphins themselves; what are they saying to each other? Probably the most famous experiment in dolphins communication research was carried out in 1965 by Jarvis Bastian of the University of California at Davis. He placed a pair of bottlenose dolphins in adjacent tanks so that they were isolated visually, yet could still hear one another. The female was taught to push paddles in order to receive a reward. The male was offered another set of identical paddles but received no training. He learned, however, to push the right paddle.

The only way the male could have obtained the information would have been from the female next door. Throughout the tests both animals were heard to emit many whistles, squeaks and clicks. Although the information must have come from the female, unfortunately the results were not conclusive proof that deliberate communication had been taken place. The male may have picked up sounds inadvertently made by the female and quite independently trained himself to use these to his own advantage a remarkable feat in itself. There have been many stories of dolphins in the wild using sophisticated communication system. It is reported that dolphin schools will avoid boats out to capture them. It has been recorded even that dolphins will steer clear of a particular type of boat simply because the shape is similar to those that are out to them harm. It was thought that somehow or other dolphins were able to tell each other 'to avoid that boat over there; one of the school was killed by a similar-shaped boat last week'.

On the other hand, dolphin schools are hauled out and killed in their hundreds by tuna fisherman around the shores of Japan.

No complicated communication system saves them. If messages carrying sophisticated information are passing to and fro, why don't they avoid the trap? Indeed, why do so many small whales get themselves beached to die of exposure in the sun? The paradox remains. What is clear is that dolphins have a varied vocal repertoire. There are squawks, whistles, squeaks, burps, groans, clicks, barks, rattles, chirps, and moans. The bewildering array of sounds can roughly by divided into two types–pulsed and unpulsed sounds. The pulsed sounds include clicks and burst pulses. The bursts may be included into chuckles, chirps and click-trains. Trains of click sound like rusty hinges being opened, or the whine of machinery, while some burst pulses have been variously described as 'raspberries' or 'Bronx cheers'. The more continuous sounds are the whistles and squeaks of frequency-modulated pure tones which may last for several seconds.

It was thought by the early researchers that the whistles are the main communication sounds. This may have been because the whistles are easier for humans to hear and to study, the clicks being mostly in the ultrasonic frequencies. But this was perhaps yet another piece of scientific mythology associated with dolphins and was brought into doubt when it was revealed that many odontocetes, including river dolphins, sperm whales and many other openwater dolphins, have not been heard to use whistles; they only emit pulsed sounds. Many of the pulsed-click sounds are used for echolocation and navigation; some, though, are thought to have social functions. Male bottlenose dolphins are harbour porpoises *Phocoena Phocoena* have known to be give pulsed 'yelps' during courtship.

Similarly Atlantic spotted dolphins *Stenella plagiodon* 'squeak' during training sessions in captivity. Frightened or distressed dolphins emit pulsed 'squeaks', which could be a type of alarm call. Aggressive 'buzzing' trains of clicks are heard when two males confront one another. What seem like exchanges of burst-pulses have been recorded between individuals in a school of Hawaiian spinner dolphins *Stenella longirostris*; so, too, between pilot whales *Globicephala melaena*, and between narwhals *Monodon monoceros*. All three species, are known to whistle. As yet, the social function of click-sounds has been title studied. There are, however, a few observations which allow some generalisations to be made about those animals that use whistles and clicks and those that use just clicks.

Heaviside's dolphin *Cephalorhynchus heavisidii*, the harbour porpoise, the finless porpoise *Neophocaena phocaenoides*, and the pigmy sperm whale *Kogia breviceps*, have pulsed sounds only and are all odontocetes that aggregate into relatively small groups of three to 20 individuals. Hawaiian spinner dolphins and bottlenose dolphins, on the other hand, often form huge schools of hundreds of animals that have a tendency to forage together. They are the whistlers. The very high frequency clicks travel less far in the water than in the relatively lower frequency whistles, and would be more suitable for sending messages between individuals of a small group. Whistles would carry better between dolphins in a large and spread-out group. There are suggestions that emotional meanings can be read into these sounds. Abrupt, loud sounds, for example, are given in aggressive situations. Sometimes these vocalisations are accompanied by non-vocal 'jaw-clapping' or 'tail-slapping'.

Intimate chuckling sounds are heard during bouts of caressing and touching. Dolphins touching, for example, make a sound much like the noise you get when rubbing a finger against a wet balloon. There are signature whistles. A dolphin will produce a whistle that seems to be its own whistle, which appears to be used as a long-range individual identifier-signal across the width of the school, to let every dolphin known where all the others are placed. In an experiment in 1974, in the USSR, two neighbouring bottlenose dolphins, linked by sound only, both produced whistles and rhythmically related clicks.

It was suggested that dolphins, like songbirds, have an opening identification portion of the call, the whistle, followed by a more complex message portion, the clicks. The birdsong parallel was continued further in the interpretation of the results of an experiment with tropical spotted dolphins *Stenella attenuata*. A male was captured and recorded. It produced almost continuous bouts of whistling which were probably alarm or distress calls. The calls were played back to the school from which it came and they fled, instantly. The same calls played to another school elicited curiosity and not flight. The captured dolphin's own school detected danger from the familiar, but frantic, call of the subject, whereas the 'strangers' were unable to appreciate the significance of the calls. Does each school, then, have its own vocabulary of calls, may be as a result of mimicry within the school? Dolphins are good mimics, as witnessed in early communication experiments, and related species, such as killer whales, are known to have distinct local dialects.

KILLER WHALES

The evidence of dialects among killer whales *Orcinus orca* was demonstrated by Dean Fisher and John Ford at the University of British Columbia in Vancouver. One of their first discoveries was that the sounds of any given killer whale pod are very stable. Unlike humpback whales, killer whales do not change their songs through time. In a variety of situations the animals make the same associated sounds over and over again. In several different pods, John Ford has been able to identify, on average, 12 distinct stereotyped or discrete calls which are exchanged between whales when spread out and foraging may be over an area of a couple of miles. It is thought that these calls are emitted in order that individuals in the pod can keep in touch with each other, although out of direct visual contact. It is now known yet whether each of the 12 sounds has a different meaning because, below the sea, it is difficult to link sounds with any particular pattern of behaviour. The Vancouver research team feel that the most likely information being communicated includes their position, individual identity, emotional or active state, and pod identity through the dialect. It looks unlikely that a more highly structured and sophisticated message, like human speech, is being exchanged. One call is given, the other in the group respond, and then they all switch to another type of call. Dialects have arisen as a result of the isolation of one pod from another. Killer whale pods are essentially extended family groups.

Once an animal is born into a pod it rarely leaves the group. In this way a pod may grow up to 50 strong, although the average

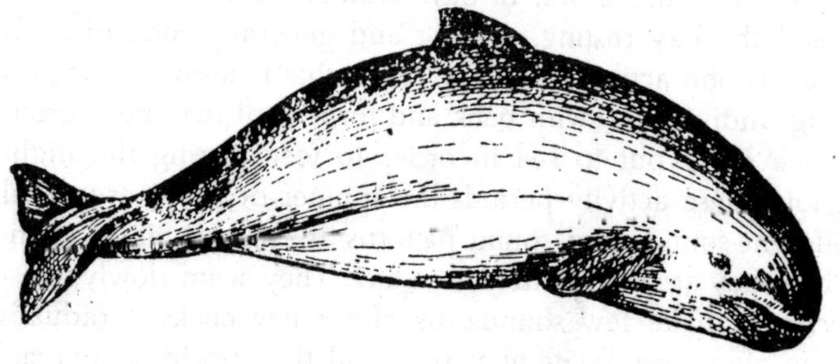

Fig. 6.9. Phocaena (Porpoise).

family groups consists of between six and 15 individuals. If an ancestral pod grows to a certain size, it is likely (although not proved) to split into two or more smaller groups, which spend progressively less time together. Initially the calls given by the groups would be similar but as time went by, probably over several decades, the dialect of each group would drift away into its own distinct sound and shape. Around Vancouver, for example, resident pods hunting in the same bays and having their own clear patrol areas tend to have very similar calls, whereas visiting migratory or transient groups can be heard to have distinct calls indicating a quite separate ancestry.

Hawaiian Spinner Dolphins

One group of dolphins whose social interactions have been studied fairly extensively are the Hawaiian spinner dolphins of the Pacific. Much of the early work was with dolphins in captivity, but increasingly today researchers prefer to work with natural, relatively undisturbed groups of animals in the wild. Dolphin watchers, Professor Ken Norris and Sharron Brownley, from the University of California at Santa Cruz, have been observing dolphins in the wild, in particular the Hawaiian spinner dolphin–easily identified by its twisting leap. They have been looking at the structure of dolphin schools, at where individuals move, how they interact socially, and listening for the sounds they make during different behaviour periods throughout the day and night. Spinner dolphins, according to Sharron Brownley, are very convenient animals to study for they do everything together.

They sleep, play, feed and travel together, and over the course of the day act in a predictable manner. At dawn a large group might arrive from night-time feeding and split into smaller groups as it gets close to the shore. In bays around the Hawaiian islands they spend the day resting, playing and generally socialising. In the late afternoon activity increases with what is known as zig-zag swimming. Individuals swim back and forth until the entire group move once more out to sea in order to feed during the night. Throughout these activity periods the spinner dolphins are vocal, with different sounds and sound patterns at different times of the day. When they are resting they are quiet. They swim slowly, close together, and make few sounds, usually a few clicks. Gradually during the afternoon excitement rises and they begin to increase the number of whistles and burst pulses. The whistles are thought to represent individual identities, while the burst pulse signals indicate to each other their emotional states, whether angry or playful. Each activity period blends with the next. The sounds then change from predominantly burst-pulses and few whistles to more whistles and fewer burst-pulses.

On the surface, activity hots up considerably as dolphins are seen leaping and twisting out of the water. Underwater they roll over each other and play, all the while emitting sounds. The more physical activity, the more noise. Sharron Brownley feels that when the animals go into zig-zag swimming it is almost as if they are trying to decide whether they are ready to go to sea to hunt. They swim slowly to the entrance of the bay and then rapidly swim back. Cape hunting dogs in Africa go through a similar ritual before hunting. Here cooperation is the key to successful hunting. Before leaving for the plains they go around twittering to each other, getting progressively more excite about the hunt and therefore more ready to hunt as a unit. The pitch of excitement rises and the level of noised rises until of a sudden they all take off and track down their evening meal. Similar kinds of prehunting calls are heard from other animals that hunt in groups, such as wolves and hyaenas. Here a dominant animal assembles the group.

In wild Hawaiian spinner dolphins schools it is difficult to identify whether an animal is male or female, let alone dominant or submissive, although researchers believe there may be group elders which shape the direction and guide the school. In other dolphin species, such as bottlenose dolphins, dominance hierarchies have been observed where a large female is dominant over the other females and sub-adults and a large male is dominant over the entire school. In Hawaiian spinner dolphins, however, when one dolphin starts calling, the rest tend to chip in too so there are periods of hectic vocalisations followed by periods of silence.

Sharron Brownley suggests that the bursts of noise may indicate how unified the group feels. When all the dolphins chime in at the proper time they know they are all ready to hunt. If some do not join the chorus then it might indicate they are not ready to go. The chorus then starts again until everybody is alert and ready. The peak period of vocal activity, then, is late afternoon. The noise level rises noticeably. The wild cacophony of sound is so incomprehensible and so full of all kinds of sounds that Ken Norris has nicknamed it the 'Yugoslavian-news-report'. Non-vocal sounds also accompany the activity. Animals leap from the water, slapping their heads and tails, and generating loud percussive noises which may help whip up excitement in the school. The leaping and spinning seems to be infectious, spreading rapidly through the group. Both visually and acoustically, leaping, slapping and whistling serve to tell where each is located and whether ready to go. Once at sea they use click for navigation and hunting.

Dolphin Whistles

Whistling associated with feeding has been observed with many different odontocete schools, both wild and captive. In dolphinaria bouts of whistling coincide with the main feeding periods. Douglas Richards at the University of Hawaii made recording of a pair of newly arrived bottlenose dolphins and found that they whistled least at night (the period of low activity for this species) and most early in the day. A year latei whistling became synchronised with routine feeding and training sessions. Dolphins riding the bow-wave of large boats or humpback whales are heard to whistle, as are individuals in unfamiliar situations, such as those stranded, captive, or otherwise isolated from the school. The level of excitement is accompanied by a comparable level of whistling.

Encounters with unusual objects or potential predators, on the other hand, will elicit silence or reduced vocal activity. When two school meet, whistling activity may increase or decrease dramatically. It is not clear why. Animals harpooned whistle continuously, as will mothers separated from babies. Captive dolphins whistle when introduced to their new tank, although they quickly settle down. Many attempts have been made to catalogue dolphin whistles. Often, several identifiable whistles are recorded together with a multitude of minor variations. It has been suggested that these animals have a graded system of vocalisations, where each sound type overlaps with another to form a series of sounds rather than discrete signals. The same kind of grade series has been described for certain monkeys and apes.

Dolphin Clicks and Echolocation

The other main category of dolphin vocalisation, pulsed sound, if often, and in most cases mainly, used for echolocation and navigation. Echolocation is sometimes considered as autocommunication or communication with self, and broadly meets the definition, used earlier, that an animal has communicated when it has transmitted information that influences a listener's behaviour. Most of the toothed whales are thought to be able to interrogate their environment with sound. Using echolocation clicks a dolphin can see with sound. By bouncing sounds off an underwater target and analysing the signal it gets back, a dolphin is able accurately to locate the object, determine whether a target is dead or alive, and if alive and potential food may be able to stun it, and sometimes kill it, with a high intensity beam of sound. That dolphins are able to echolocate was revealed in 1942, and although since then experiments have mushroomed, we still know very little of the nature and function of a dolphin's echolocation system. One of the workers who has been at the centre of dolphin research, right from the early days, Ken Norris of the University of California at Santa Cruz. His pioneering work was with captive animals in dolphinaria or marine circuses; indeed, the only way to fund serious research was to teach dolphins new tricks. One of his first tests was for a TV show.

Ken Norris was allowed the use of a dolphin on the understanding that it could be televised, and therefore also it had

Fig. 6.10. Bottle-nose dolphin.

to be entertaining. The researchers noticed that when a dolphin swam between a pair of hydrophones placed in the tank the sounds would increase in volume if it swam directly towards one hydrophone. Then as it turned and swam towards the other, the sounds would appear to come up in the second hydrophone. Clearly sound was being emitted in front of the animal, but proof was needed that it was being used for echolocation. It was necessary, therefore, to blindfold a dolphin in order to determine whether it could use sound alone to get about in it tank. The researchers had many failures for its is very difficult to tie anything on to make anything sticks to a dolphin. The solution to the problem was a pair of rubber suction cups, one placed over each of the dolphin's eyes. The test animal, a bottle nose dolphin named Kathy, swam of across the tank blindfolded as if nothing was strong. With ease she picked her away through a maze of poles without every touching one, and was able to locate small objects on the far side of her ten meter tank. Small pieces of fish dropped right next to a barrier in the maze would be scooped up without touching the obstacle.

Another test with Kathy was a discrimination test. She was invited to distinguish between a horse capsule filled with water and a piece off fish about the same size. Kathy took the fish every time. Dolphins clearly can use a sense other than sight to navigate and locate food. As dolphins have all but lost their sense of smell, and extra-sensory perception has zero scientific credibility, sound was considered to be the likely candidate. One of the problems, though, is understanding how dolphins make these sounds. They don't have vocal chords and are rarely seen to blow bubbles when vocalising. This is a controversial area of research. One school of thought has held that the sounds are produced in the larynx just as in the other mammals, but echolocation clicks seem to emanate from the forehead rather than the throat. Indeed, if microphones are placed around the head of a 'clicking' dolphin the sounds can be triangulated to deep in the forehead, at the back of the nostrils. If probes are placed in the muscles of the larynx and nostrils has been found that during sound production the larynx is quiescent while valves in the nostrils show muscle activity. Ultrasound scans have given similar results.

At Boston University, R. Stuart Mackay and H.M. Liaw projected narrow beams of low-intensity ultrasound at a frequency too high for the dolphins to hear and were able to identify the structures that moved during sound production. The apparatus was a modified foetal heart monitor of a type found in most maternity hospitals. The observers were able to see the nasal plug and the vestibular, nasofrontal and premaxillary air sacs vibrate with clicking or buzz sounds. Nasal diverticula on the right side vibrated all the time when clicks are produced while the left nasal diverticula vibrated only some of the time. The vestibular sac inflated as the clicking sound was made, probably as a resonator. The nasal plugs were thought to be the site of the original sound production as air moved upward, presumably from the lungs. When the blowhole is closed, air recycles to the vestibular sac.

Clicks seem to originate in the right diverticula. Whistles are thought to be generated in much the same way as human whistles except that they are 'blown' internally. Dolphins have a pair of nostrils inside the head which come together under a single blow-hole. The and ancestors of these animals probably possessed paired external nostrils, just as most mammals do today, but as dolphins evolved into divers they needed a way of storing air if they were to make and use sound underwater. They developed a covering over the nostrils, blow hole, with a complicated series of nasal sacs and valves below. Air is breathed in at the surface, and on diving, the blowhole is closed.

The air trapped in the dolphin's respiratory system can then be passed across the valves in the nasal sacs to produce sounds and then recycled through a complicated series piping to be used over and over again as the dolphin echolocates underwater. The click sounds are thought to leave the dolphin's body through the forehead. At the front of the forehead is a large fatty body known as the melon which focuses the sound, much as an optical lens focuses light, about a metre in front of the animals's head. Donald Malins and Usha Varanasi, of Seattle University have found a concentration of unusual lipids, made of isovaleric acid, at the centre of the dolphins forehead. The three dimensional arrangement of these small molecules, rarely found in the lipids of other animals, suggest to the researchers that the area is, indeed, a 'sound lens'. When the sound returns having bounced off a target, it does not enter an ear canal but is picked up by the thin bone of the lower jaw as a vibration, travels along fatty tissue in the hollow lower jaw, and thereby is transferred directly to the middle ear. Sound

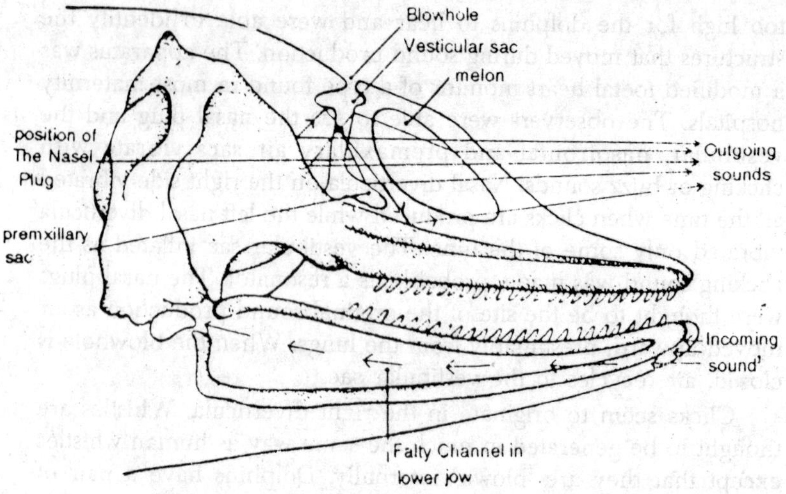

Fig. 6.11. Simplified section of dolphin head showing structures associated with sound production and detection.

production, transmission, and reception is optimised for life underwater. The rapidity with which the clicks are emitted means that humans cannot hear individual packets of sound, rather we hear trains of pulses much like a creaking old door hinge closing.

One of the remarkable things about dolphin echolocation is the rapidity with which the ear and brain process these clicks. A click-train may be made from up to 700 units of sound per second. A dolphin is capable of mentally separating these units, listening to the individual echoes, and decoding information while interrogating a target. In the human ear the sounds would fuse together in our minds at 20-30 clicks per second. The sound is also known to enter solid structures. In experiments, dolphins have been able to tell the difference between a copper and aluminium plate painted the same colour. They also have the ability to distinguish a hollow aluminium tube from a solid one, although both tubes looked identical from the outside.

Perhaps one of the most interesting aspects of dolphin research centres on the sound beam itself. Modern dolphins have narrow beams of sound; the bottlenose dolphin, for example, has a beam width of about 9°, a thin pencil-beam of sound. The Indus river dolphin *Platanista indi*, a more primitive species, emits a 65° beam. Part of a evolution of these creatures might have been a narrowing of the beam width, giving a greater distance penetration for the same amount of energy. The concentration of energy in narrow-beam dolphins has become so intense that the prey being detected has become affected by it, leading to an entirely novel way of catching lunch. Together with Bertel Mohl, of Aarhus University, Denmark, Ken Norris has been pursuing the idea that some of the toothed whales may have the capacity to debilitate their prey with sound. The sound put out may be so intense that the prey may be killed, or at least immobilised to prevent escape. The idea is not new. Drs. V.M. Bel'Kovitch and Yablokov, in the USSR, calculated that dolphin sounds should have enough energy to-incapacitate prey. Further support came from the work of Dr. A.A. Berzin with sperm whales. Berzin looked at a number of sperm whales caught by the whale fishing industry.

Strangely, some of them came up with what looked like congenital deformities of the lower jaw. The lower jaw was curved so the animal was unable to close its jaws together to catch prey. In the stomachs, however, were plenty of squid and the whales looked otherwise perfectly healthy. Squid swim much faster than sperm whales, anyway, so it was doubly surprising that these enormous animals, the size of an omnibus, were able to take the ton or so of squid per day needed to survive. Berzin concluded that the jaws were not essential for feeding and that sound was being used to stun the squid. In Hawaii, Whitlow W.Au and A.E. Murchison measured the intensities of sounds emitted by a bottle nose dolphin that was asked to carry out extreme-distance discrimination tests. A sphere, about the size of a tangerine, was placed about 143 metres from the dolphin. The animal was able to locate the sphere, but interestingly, when the echolocation sounds being produced were measured, they proved to be many orders of magnitude higher than any sounds recorded from a dolphin before. Indeed, they were close to the finite limit sound, that is, the limit at which any more energy put into the water would simply turn to heat. With that kind of sound level Ken Norris asked the question–will those sounds kill prey? Tests with man made sound beams showed that fish could be killed with a high intensity sound beam.

At the Marine Biological Laboratories at Plymouth, large squid could be killed quite rapidly by sound. In theory, then, dolphins could kill with sound, but the next question was whether they used this weapon. Norris considered that it would be odd for a creature to have evolved such a weapon if it didn't use it. Further tests were carried out with dolphins in capacity. Live fish were placed in a tank with three Hawaiian spinner dolphins, and sure enough after the short while the dolphins began to spray the fish with sound. The fish used were much larger than the dolphin's normal prey so it was not expected that the dolphin's would kill them. Norris and his colleagues, though, watched for any signs of debilitation.

The dolphins continued to direct echolocation skills at the first school for over an hour. A dolphin would make a run at the school attempting to put the fish right on the tip of its beak. The fish school would split, head for the tail of the dolphin and reform. Slowly, though, the fish school became depolarised. The researcher noticed after a while the fish were not all facing in the same direction. Then individuals began to wander away from the school, seeming totally disorientated. Similar observations have been made in the wild. Striped dolphins *stenella coerulealba*, for example, have been seen to circle an anchovy school, spray them with sound, and then cut through the school shovelling fish into their jaws at will. The anchovies do not attempt to escape. It is possible that the fish eventually suffer from a build-up of waste products in the blood due to their strenuous efforts to escape but an interesting observation made by a biologist on a fishing boat off Vancouver does not appear to support that. A large salmon was clearly visible in the water below the boat.

As the researcher watched, the salmon suddenly stopped swimming and directly behind a group of killer whales approached. One of the whales scooped up the salmon and swam on. The salmon made no attempt to escape. Another piece of anecdotal evidence is that many scuba drivers in dolphin circuses have felt a light touching sensation on the backs of the necks which they have attributed to the echolocation activity of their charges. If the interpretation is correct, dolphins are clearly formidable killers, but one problem individuals in a large dolphin school might have is the danger of zapping another dolphin. In a school of actively feeding dolphins it would be very easy for one animal seriously to upset another, and dolphins are known to get angry with each other. To look for signs of 'echolocation manners', Ken Norris watched and listened carefully to three individuals in a tank. In more than a hundred crossing of one dolphin with its fellows, it never once echolocated them.

As a dolphin crossed in front, the actively echolocating animal switched its gear off, turned its head away, and then switched back on again. With such a devastating weapon in the dolphin's head, it is important that 'echolocation manners' are a part of the social organisation of a dolphin school. Large schools of dolphins may consist of hundreds or even thousands of animals which hunt in a 'line-abreast' formation. In this way a broad expanse of ocean can be echolocated in the search for food. Sound is thought to be used also in locating good food-gathering areas. Dolphins, it seems, can listen to the underwater background noise associated with the increased number of organisms at submarine escarpment and sea-mounts. There is also evidence that small schools of bottlenose dolphins sometimes detect and follow the sounds being made by pilot whale schools.

The pilot whales, with their longer diving times, appear to be better at finding food and the dolphins take advantage of this superior food-finding ability. Having located a school of fish, a large school of dolphins will change its shape during the attack although they will always dive in synchrony. Sound signals must keep the school in step. Killer whales have been seen to work together in a similar way to capture prey. Observers of harpooned baleen whales off Australia have described pod members hanging on to lips, flukes, and lying on the blowhole in attempts to immobilise the whales. On other occasions killers will encircle prey animals like seals or walruses, coordinating their moves by sound. When approached by killer whales, other cetaceans often become silent, flee quickly or 'spy-hop', that is they rise out of water and search the surface.

7

Social Behaviour in Wild Animals

About 37 million yeas ago, when primitive fissiped carnivores were evolving from their miascid predecessors, some of them forsook their way of life on land and went to sea to find food. This move must have worked out well, because many of the descendants of these pioneers–the seals, sea lions and walruses–are still making a living in the sea today. No doubt, descendants of the carnivores that remained on land fared well, too. Because land and sea are such different environments, we might expect the two animals groups to have diverged in many ways since their separation. Indeed, it should be interesting they have in common, an ancestor, and because of the obvious difference in the environments in which they live. In this chapter, some differences and similarities aquatic carnivores like between seals and their terrestrial counterparts with respect to gross morphology, feeding and predatory behaviour, reproductive behaviour, and various aspects of social life have been discussed. The similarities between aquatic carnivores and land carnivores were obvious to Laymen and early scientists. Fifteenth and 16th century sailors described the seals they saw as marine carnivores and called them *"sea lions," "sea bears,"* or *"lobos de mar"* (*sea wolves).*

Linnaeus and several later taxonomists classified seals, sea lions, and walruses in the order Carnivora. However, as early as 1811, there was an attempt to place these animals in a separate order (*Illiger,* 1811). No a days, many investigators class these marine mammals in their own separate order, the Pinnipedia. However, classification of these animals if far from settled. Some maintain that the new order is inappropriate and that pinnipeds should be considered merely a suborder of the Carnivora. Other argue that the pinnipeds can be classified into two living families (walruses are placed in the same family as sea lions) under the super family, Canoidea, in the order Carnivora. The earliest pinnipeds probably entered the sea in one of the following areas: the Arctic Basin, the northwest coast of North America, or the Tethyan-Mediterranean area. It is not clear whether seals and sea lion (including the walrus) were already differentiated at this time. The oldest remains of pinnipeds are from the Miocene era, a time when they were already well adopted for aquatic life and when seals and sea lions were already distinguishable.

One opinion holds that all pinnipeds derived from canoid or dog-bear stock. An opposing view is that sea lions and the walrus derived from *ursine stock, and* seals derived from *lutrine stock.* This question is far from settled. Nevertheless, there is agreement that all pinnipeds are more closely related phylogenetically to members of the superfamily Canoidea of the order Carnivora the dogs, raccoons bears weasels and the like–than to the feloid carnivores. After their entrance in to the water, the distribution and diversity of pinnipeds were influenced by geomorphic and climatic barriers, distance between land falls, ocean currents, and water temperature. *Davis* hypothesizes that pinnipeds are, and always have been tied to a cold water environment. He argues very persuasively that the distribution and differentiation of present day northern pinnipeds reflects periods of expansion and contraction of sea and glacial ice that occurred during the Pleistocene era.

Scheffer points out that "as local shore lines and islands rose and fell and glacial barriers came and went, pinnipeds moved back and forth in order to maintain favourable breeding grounds along the edge of the sea." On land, many contemporary carnivores were being forced into extinction by periodic fires, floods, droughts and dust storms. *Scheffer* suggests that the early pinnipeds moved out along the shores of continents from island to island and later to polar ice fields. The rate or evolution was faster at the frontiers of advancing lines where immigration was unidirectional. The primitive populations were small and the members did not wander far, had no well developed homing instinct, were not migratory, and did not need to rendezvous in special places in order to find mates. Polygyny had not developed in any of the pinnipeds stocks. The animals were perhaps smaller than most recent pinnipeds had less fat, lived in more temperate waters, and perhaps made crude hens in beach grasses and among boulders.

The advancing pinnipeds met, from sea birds, and cetaceans, little competition for food. They met no competition for breeding room nor do they often today. Pinnipeds early lost the habit of feeding on beach organisms such as crabs, mussels, periwinkles and blennies. As each generic stock became isolated, it was transformed in response to the local physical and biotic environment and adapted to its peculiar niche. Among species frequenting isolated bays, gulfs, islands, inland seas, or lakes, there has been and still is, rapid evolution in the last few centuries, the effect of commercial sealing upon island and continental pinniped population has been catastrophic. Millions of seals have been slaughtered and entire breeding populations wiped out.

Fig. 7.1. Lutra (Otter).

It is doubtful that some species will ever recover genetically from undergoing such a severe population "*bottleneck*". In one recently exploited species, absolutely no polymorphic variation was found in 21 blood proteins fro 159 seals representing five rookeries! In this regard some land carnivore populations have suffered a similar fate at the hands of man-for example, the wolf. The increasing range of human populations is changing the distribution of both land carnivores and pinnipeds. Inspite of the fact the pinnipeds are marine, all species have retained an attachment to land.

In this respect, they differ from the Cetacea, descendants of primitive ungulate ancestor, who have become completely aquatic. All pinnipeds give birth to their youngs on land or ice, and some species spend the majority of their time resign out of the water. Pinnipeds have exploited the marine niche for food. A few carnivores, such as the sea otter, *Enhydra lutris,* and the polar bear, *Thalarctos maritimus,* also obtain food from the sea, but they have not undergone the same degree of morphological transformation to aquatic living as the pinnipeds. The anatomical and behavioural adaptations of the pinnipeds to life in the sa these adaptations influence their behaviour on land, it is worth reviewing some of the major ones.

Morphological Comparisons

Size

There are several morphological difference between pinnipeds and land carnivores simply reflecting the different environments in which they live. The average body size of pinnipeds is greater than that of carnivores. The smallest pinniped, the ringed seal,

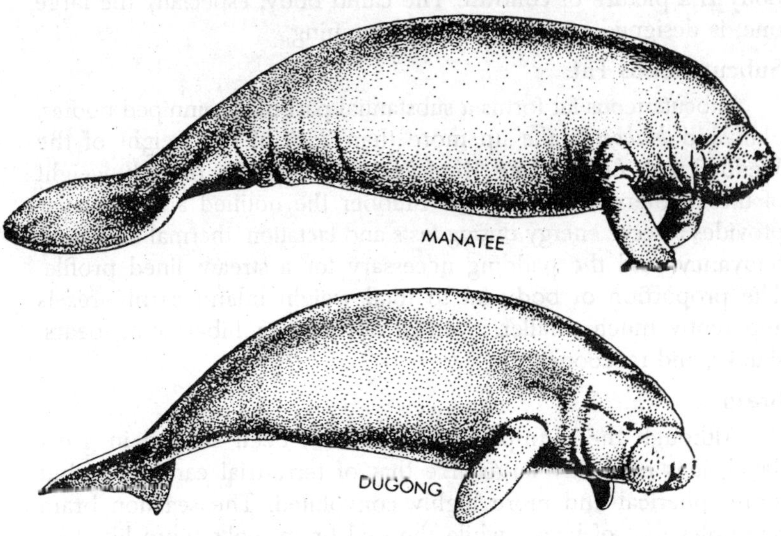

Fig. 7.2. Two Sirenian representatives.

Pusa hispida, weighs 90 kg. The largest pinniped, the southern elephant sea, *Mirounga leonina,* weighs 771 kg, about 3629 kg and is 650 cm long. Several seal species are larger than the largest land carnivore, the grizzly bear, *Ursus arctos,* which weighs. Many land carnivores such as the foxes of Africa weigh less than 4 kg.. The Fennec fox, *Fennecus zerda,* weighs less than 1 kg! The greater body size of environment. Large body size is better fro hear retention in the cold sea. Moreover, since the water medium gives more support than air, large size could more easily evolve in the sea than on land.

Shape

Aquatic adaptation favouring swimming and diving have been observed in pinnipeds Body shape is streamlined to reduce drag. The external ears are reduced or absent, External genitalia and mammary tests are drawn into the body, Limbs are enclosed within the body and extremities are flattened the tail is short and the head is flattened, and the eyes are situated well forward. The neck, in particular, is thick and muscular and considerably more flexible than in most large carnivore. The skin also shows adaptation to a water environment, and the hair is flattened. Pinnipeds never groom the pelage with the mouth or tongue. The terrestrial carnivore body is a picture of contrast. The canid body, especially the large one, is designed for long distances running.

Subcutaneous Fat

Subcutaneous fat forms a substantial part of all pinniped bodies. Skin and *blubber* make up more than 25% of the weight of the Weddell seal, *Leptonychotes Weddelli* and almost 50% of body weight of the southern elephant seal. Blubber the notified adipore layer provides reserve energy during fasts and lactation, thermal insulation buoyancy, and the padding necessary for a stream lined profile. The proportion of body fat to total weight inland carnivores is apparently much smaller, even in dormant or hibernating bears, skunks, and raccoons.

Brain

Although the pinniped brain has not been studied in great detail, it is evidently large, like that of terrestrial carnivores, but more spherical and more highly convoluted. The sea lion brain resembles that of bears, while the seal brain looks more like that of cats ad dogs. Compared to a dog's brain, the cerebellum is large, probably because of the increased coordination demanded in swimming. The auditory nerve is large and the olfactory lobes are reduced. Pinniped eyes (the walrus excepted) are larger than those of land carnivores, and they function well at low levels of illumination. Seal vision is excellent both in water and on land.

Teeth

Pinnipeds have fewer and more uniform teeth than most land carnivores (except for the walrus, a special case once again). All species have the pronounced upper and lower canines but lack the carnassial cusps characteristic of land carnivores. The postcanines in most species are rudimentary pegs functioning to hold prey that is swallowed whole. Variation in pinniped teeth reflects the type of flesh the animals feed on and as in many mammals, it is a useful taxonomic indicator. Generally, the teeth, mouth, jaw, and associated structures of pinnipeds are designed for grasping tearing and swallowing prey whole or in chunks, rather than for chewing shearing or crushing large bones.

In contrast with land carnivores, the deciduous teeth or milk teeth of pinnipeds disappear before or soon after birth. Several physiological adjustments to aquatic life have bee made on pinnipeds that set them apart from land carnivores. These adaptations in respiration circulation renal physiology, and the like are important and interesting, but it would be too much of digression to cover these topics here.

Prey, Food Habits, and Feeding Behaviour

Pinnipeds, deviated from land carnivores in form and function because of adaptations that took place as a result of seeking food in the seal. It is therefore most important to compare these two animals groups from the point of view of hunting and feeding habits, and in relation to their respective prey.

Prey

All carnivores including pinnipeds and land carnivores are meat eaters, but only the marine mammal eat flesh exclusively. Seals and sea lions feed mainly on crustacea, molluscs, fish penguins, and other sea birds. Although diet varies with the species, the majority of pinnipeds feed on a variety of prey. The Alaska fur seal, *Callorhinus ursinus,* is the best studied pinniped fro the point view of food habits best studied pinniped from the point of view of food habits. It catholic diet is representative of the wide range of fishes and squid eaten by the Otariids (the earred seals or sea lions and fur seals, as opposed to the Phocids, the earless, true seals). Alaska fur seals feed on more than 30 kinds of marine organisms: at leas 27 species of fishes. 1 species of octopus, and 5 species of squid.

The most common prey of fur seals are northern anchovy, *Engraulis mordax,* squid and Pacific herring. (*Clupea harengus).* Various rockfish, Pacific hake, *Merluccious productus,* Pacific saury, *Cololabis saira,* salmon, *Oncorhynchus* spp., and American shad, *Alosa sapidissimia,* are eaten in lesser quantities. Hermit crabs, amphipods, and various dividing sea birds (e.g., Rhinocerous Auklet, *Cerorhinca monocerata,* Redthroated Loon, *Gavia stellata,* and Best Petrel, *Oceanodroma leucorhoa* beali) are eaten occasionally, but they seem to play only a minor role in the seals' diet. Fur seals, like most other pinnipeds, are *opportunistic feeders,* What is eaten depends on the seasonal distribution and the abundance of prey. Since fur seals commonly feed on schooling fishes, stomach contents of several animals collected in the same area often contain only one type of fish. The Steller sea lion, *Eumetopias jubata,* and the California sea lion, *Zalophus californianus,* eat similar fishes and squid, especially where the feeding range of the three species overlaps. Steller males in the Bering Sea and off Alaskan islands supplement their diet with Alaska fur seal pups. Most true seals, such as the harbor seal, *Phoca vitulina,* and the grey seal, *Halichoerus grypus,* are mainly fish eaters, but they occasionally fed on an assortment of crustacea, octopus eels and molluscs.

The northern elephant seal, *Mirouna angustirostirs,* is a deep dividing seal that feeds seal primarily on squid skates, rays, small sharks and ratfish *Hydrolages collier.* The small seal, *Pusa Hispida,* which inhabits the circumpolar Arctic coasts, feeds on up to 72 different species of small pelagic amphipods, euphausians, and other crustacea, as well as on small fishes. Only two of the 32 different pinniped species are rather specialized feeders that exploit one type of prey species almost exclusively. The Crabeater, *Lobodon carcinophagus,* feeds on krill, small shrimp like animals that it catches in quantity and strains from the water through its multicusped cheek teeth, much as a mysticete whale sieves krill through its baleen. The walrus, *Odobenus rosmarus,* feeds primarily on three genera of bivalve molluscs, *Mya, Saxicava,* and *Cardium,* for which it forages in shallow coastal waters less than 40 fathoms deep. Bivalves on the sea bottom are examined and sorted by the lips and whiskers; feet and fleshy parts are torn off and swallowed whole or sucked out. But when molluscs are scarce, the walrus may eat young ringed seals, bearded, seals, *Erignathus* apparently feeds on Cetacea.

Occasionally a narwhal, *Monodon monoceros,* or a beluga, *Delphinapterus lcucas* is eaten. However, it is not known whether walruses actually kill them or feed on corpses. According to *Mech* "probably every kind of backboned animals that lives in the range of the wolf has been eaten by the wolf". Wolves, *Canis lupus,* eat mice, mink muskrats, squirrels rabbits, various birds fishes lizards snakes , grasshoppers, earthworms, and berries. But predation on small animals plays only a minor role in the wolf's diet. The main prey of the wolf, like those of the African wild dog. *Lycaon pictus,* are large animals. The wolf's primary prey in the United States and Canada are : white tailed deer, mule deer moose caribou elk Dall sheep, bighorn sheep and beaver. Wild dog kill mostly Thomson's gazelles juvenile wildebeests and Grant's gazelles and occasionally, warthogs and zebras.

African foxes feed mostly on rodent and small reptiles birds's eggs, insects and vegetables, matter, jackals eat similar foods in addition to small mammals and carrion. Additional information on other canids can be found in *Fox.* Of all the land carnivores, bears have the most diverse tastes. The grizzly, before man virtually annihilated it, was an omnivorous opportunist *part excellence.* The grizzly ate almost anything and everything that was available. This long list included: meat, fresh or putrid from whales, water birds, fishes, elk, deer, antelope gophers lizards, frogs and domestic livestock: a wider variety of plant materials than herbivorous animals ate, some of which were various berries, clovers, nuts, wheat corn, potatoes tree bark and various bulbs and assorted foods like honey, ants and their larvae, yellow jacket nets, and mushrooms. Polar bears, *Thalarctos maritimus,* have a much more restricted diet. Although they consume some vegetation and carrion during the summer, they feed primarily on the ringed seal during most of the year.

Size of Prey

Among large carnivores there is a remarkable thing about their feeding habits is that the prey is often larger than the predator. Wild dogs, whose average weight does not exceed 18 kg, bring down impala and reedbucks that are double or triple their size and zebras that are even larger. Wolves weighing 36 to 45 kg kill moose and bison that may weigh over 500 kg. These canids overcome large prey by hunting in packs that range from a few animals up to a score or more of them. One dog usually selects a quarry from a retreating herd of gazelles, and the other dogs follow it . The quarry is coursed for 1 to 3 km until it becomes exhausted. The lead dog catches up to it and grabs it or bowls it over. Once overtaken, they prey is fallen on by all the following dogs, and it is dismembered an eaten with great dispatch. Wolves bring down deer or moose in a similar way. The first wolf to catch up to the fleeing prey attempts to grab hold of the rump, flanks, necks or nose. Once the prey is down, other attack from every side.

Although the technique my vary with prey species, the strategy of the individual quarry and the composition of the hunting pack, it is clear that these pack hunters get help from each other in getting their food. Lone wolves or even small packs are at a disadvantage in bringing down large prey. In black backed jackals, *Canis mesomelas* two hunters are more than four times as successful as one in bringing down gazelle fawns as observed by Wyman. In pinnipeds the size of prey eaten varies greatly like land carnivore, but except for the dobious possibility that walruses kill small whales, pinnipeds always kill and eat animals that are much smaller than themselves. Prey size determined how alaska fur seals, Steller sea lions, and harbor seals eat their food. Small fishes less than 30 cm long, such as anchovy, herring saury, or squid are consumed whole under water. Except for very small food items like lanternfish, the prey is swallowed head first. Larger prey measuring more than 30 cm long, such as hake, rockfish, or salmon, is brought to the surface, grasped by the head shaken violently and reduced to chunks that are swallowed piecemeal.

As fur seal get older and larger, large fishes up to 92 cm make up an increasingly high proportion of their diet. Thus the size of the prey increases with the growing age of fish seas. In most of pinnipeds the hunting is done in groups, but, unlike the case with canids, grouping does not enable them to exploit larger animals. Rather, grouping seems to make it easier for them to exploit large aggregations of small prey. *Fiscus* and *Baines* saw Steller sea lions leaving their hauling grounds in compact groups of several hundred to several thousand animals. They swam out to feeding areas, where they dispersed into smaller groups of less than 50 animals containing both sexes and mixed sizes of animals. Massings of this type fed on large of sea lions feed or squid.

Casual observations suggest that groups of sea lions feed more efficiently on schools of fish because cooperation among the hunters enables them to herd and control the movements of the school. When large fish schools are absent, sea lions feed singly or in small groups of two to five animals. In some cases the hunting is performed in a very special way before Steller sea lion females leave the rookery and their pups to go to sea and feed, they may engage in activities that resemble and that may have an analogous function to, the prehunt greeting ceremony of wild dogs.

Several females may gather at water's edge, where they mill about restlessly in close contact with each other. They may vocalize repeatedly and engage in what appears to be low intensity aggressive behaviour before swimming off as a group to feed. In both species, these activities may reinforce group cohesion and unity as well as synchronize the time of departure. Similar observation have also been made in canids and pinnipeds. They may travel several kilometers to get a meal. Van *Lawick-Goodall* and *van Lawick-Goodall* report that wild dogs did not start their firs chase until they were 8 km from where they started. The chase itself covered a distance of $5^{1/2}$ km. Wolves may pursue moose or caribou for up to 5 to 8 km but usually give up sooner. They may range up to 32 km from their den. Steller sea lions and California sea lions travel much further from their rookeries or hauling grounds to feed. The former have been seen feeding as far as 112 to 136 kms from land and the latter as far as 62 kms from land.

Usually, only small groups of less than 30 individuals are seen this far from land. Large groups of sea lions (100 or more) are seldom seen feeding more than 16 to 24 km from a hauling ground or rookery. It would be it interesting to compare land carnivores and pinnipeds on the success of the respective hunting and fishing expeditions, but the difficulties inherent in observing seals feeding in the water makes this impossible at present. However, some indirect evidence is pertinent. Although both pinnipeds and the larger land carnivores are nomadic, the former travel greater distances in their annual feeding migrations. The record holder is the Alaska fur seal, which migrates from its breeding grounds in the Bering Sea to its winter feeding grounds along the shores of the eastern and western Pacific as far south as San Diego, California, and Hokkaido, Japan, a distance of some 5000 km. Wolves may follow caribou on their annual migrations.

Times of Hunting

Wild dogs and Wolves, *Ceron alpinus,* prefer to hunt in early morning or early afternoon or evening, although hunts occur on moonlight nights. Wolves an bears may hunt by day or night. Most of the smaller foxes are nocturnal feeders. Fur seals and sea lions are primarily night and early morning feeders. Stomach contents of animals collected at various hours after sunrise show decreasing amounts of food. However, in some areas where large schools of fish are present, daytime feeding also occurs. Typically the midday hours are spent resting, and feeding activities gradually resume in late afternoon. The harbor seal is a daytime feeder.

Amount of Food

The amount of food varies for animal to animal. Canids eat great amounts of food in a single feeding bout. A wolf may gorge 9 kg of meat in single meal; this may represent 205 of its body weight. However, estimates of little more than half that amount

Fig. 7.3. A–Phoca (Seal); B–Odobenus (Walrus).

per day per wolf may be more representative. *Estes* and *Goddard* estimate that wild dogs average 2.72 kg of meat per day, or approximately 6.7% of their body weight. But these canids may have to go without eating for several days at a time. Five to 7 days fasts have been recorded in wolves, and one wolf went for 17 days without food. It is common knowledge that domestic dogs can go for several days without eating, particularly when sexual activity is probable. Grizzlies and black bears in the more northerly parts of North America may go dormant for varying periods of time during the winter go dormant for varying periods of time during the winter when food is scarce. In general all of the land carnivores will feed regularly if possible. Even bears will actively feed through out the year when food is available.

Fasting

Among aquatic a carnival mammals the pinnipeds feed daily except at certain times of the year, usually the breeding season, when they undergo long fasts. Thus, feasts and famine are even more extreme in their annual cycle. when they are feeding fur seals and sea lions ingest less food relative to body weight than the canids mentioned above.

Spaulding estimates that fur seal, seal lion, and harbor, seal food requirements range from 2 to 11% of body weight per day with 6% being the average. During the breeding season, fur seal and sea lion males fast for approximately 6 weeks; harbor seals do not appear to fast during this time. The record for long fasts in pinnipeds goes to male northern elephant seals. They may go without food for 3 months. Throughout this time they are actively fighting and attempting to copulate, and they may lose over 450 kgs. Females of this species fast for an average of 34 days during the period in which they give birth and nurse their pups daily for 28 days. A female's total body weight maybe reduced by almost 50% during this fast. During the nonbreeding season, males and females fast again for approximately 30 days while undergoing the annual molt.

Specific Feeing

Most canids and ursids scavenge for food at some time, Perhaps bears exploit carrion more than any other land carivore. Carrion was a major part of the California grizzly's diet more than 100 years age, before kept the California coast clean of pinniped and whale carcasses that repeatedly washed ashore. There are numerous reports of 12 to 1! bears feeding at the same time on a single whale carcass. Many of the larger land carnivores typically desert remains of a kill and cache it away and return to et from it later. Polar bears are an exception to this rule. They do not cache ringed seal carcasses that they kill, despite the fact that they may only feed on the blubber and leave the meat behind. In marked contrast to these feeding habits of land carnivores, pinnipeds do not cache food, nor do they scavenge although the walrus may be an exception to this rule. Pinnipeds consume most of their prey entire and discard only the heads of large fish, particularly rockfish.

Predation

One drawback that pinnipeds experience in their marine habitat to which the larger land carnivores are relatively immune is predation. In carving their niche in the sea, pinnipeds became exposed to large predators who were also adapted to aquatic life and more formidable than themselves. At sea, the main predators on pinnipeds are large sharks, particularly the great white shark, *Carcharodon carcharias,* and the killer whale, *Orcinus orca.*

All pinniped species living in the Pacific are preyed on by these animals. The degree of predation is unknown. When pinnipeds return to land or ice to give birth and nurse their young, they are vulnerable to predation by large land carnivores such as polar bears and grizzly bears, perhaps wolves and of course, man. Small carnivores such as foxes and skuas, gulls, and hawks, may prey on their young. Being adapted primarily for locomotion in water, they are no match defending themselves against large land predator. Thus, it would appear that predator pressure was important in causing pinnipeds to breed in remote sanctuaries free from these predators; that is, offshore, islands or rocks, sandbars and ice floes, and offshore breeding areas are limited.

Social Behaviour and Reproduction

It is likely that pinnipeds were forced to share limited breeding areas, and this led to the high degree of sociality and the large social gatherings that we find in pinnipeds today. In terms of the sheer number of animals that congregate together, the pinnipeds are much more social than any land carnivore. As many as 2 million Alaska fur seals can be found every summer on two tiny Pribilof Islands in the Bering Sea. Three thousand northern elephant seals maybe found packed together tightly on one beach during the peak of the breeding season. Two thousand California sea lions may be found sleeping together in close contact during the nonbreeding season. The largest wolf or wild dog pecks are miniscule in comparison. Pinnipeds are among the most *polygynous* mammals.

Apparently, this mode of life developed very early in their history, and it was closely tied to their amphibious habits. When females began clustering in time and space, the conditions became ideal for promoting male-male competition, *polygyny,* and sexual dimorphism. Males must have competed to mate with as many females as possible. Those who kept other males away from areas containing females sited more pups.

In their short breeding seasons on traditional rookeries, fur seals and sea lions developed a social structure characterized by territoriality among males. Males without territories do not copulate, or they copulate only rarely; those who secure the most well placed territories–territories containing the most estrous females–do most of the breeding. Another system that developed was for males to dominate other males and thus gain access to more estrous females wherever, they were situated. This was the strategy adopted by northern elephant seal males. They fight for social status in a dominance hierarchy. The highest ranking male or males depending on the number of females present–locate themselves nearest the females and keep all other away. One or a few of the highest ranking males nor breed at all. One male may dominate a haem for several years and inseminate more than 200 females. Grey seal society is similar to that of elephant seals. Thus, in most of the well studied pinnipeds, males are *polygynous* and are either territorial or exhibit a dominance hierarchy during the breeding season. This is a reflection of the manner in which the largest and strongest males go about monopolizing a large number of females.

Since size and physical aggression convey such a great reproductive advantage to males, these traits have become increasingly elaborated in time. Sexual dimorphism is greater in some seals and sea lions than in other mammals. For example, the Alaska fur seal male is five time larger than the female. Among terrestrial carnivores the Canids are "*social* " in quite a different way from pinnipeds. Social organization is characterized by seasonal or permanent pair bonds in foxes and jackals, and by cooperative hunting and group life in the wolf, dhole, and wild dogs. Black bears are essentially solitary, but female cub associations are long lasting, and strong and but female cubs inherit their mother's feeding territories. Wolf packs are organized into separate male and female social hierarchies.

Although the female hierarchy is more ambiguous, it is clearly more pronounced than the size-related dominance observed in female elephant seals. Hunting dog packs have similar hierarchies, particularly during the breeding season, but it is in the female hierarchy that is most evident. In wolves and wild dogs, the dominant breeding female may prevent subordinate females from breeding or may drive them out of the pack and kill their pups. Dominant females may also enlist the help of males, and they may prevent other adults from feeding the subordinate female or her young.

Territorial Behaviour

Although many canids and ursids are territorial in the sense that they defend a feeding or home range against other individuals or groups of individuals, this behaviours is quite different from the defense of rigid individual territories by fur seal males and Steller sea lion males against conspecific males during the breeding season.

8

Lekking in Fishes and Birds

Lekking is the temporary aggregation of sexually active males for reproduction. In the typical case, males gather on a lekking ground or arena. There each male occupies a territory, or court, from which he displays to females and interacts with other males. The aggregation of males on the lek is visited by females, singly or *en masse*, who select the males with whom they mate. Once mating is accomplished, the females leave the lek. Clusters of males holding permanent, all-purpose territories, even if visited there by females for reproductive purposes, are not considered by us to constitute a lek. In a more comprehensive treatment of reproductive adaptations, their relationship to lek systems would have to be considered. In the context of this paper, such consideration would be too great a digression.

Some colleagues have suggested to us that the use of the term *colonial breeding* would be more appropriate in such a view. We disagreed, insofar as our aim is to make explicit comparisons between lek systems in birds and fishes. That term has two significantly different applications in avian literature. The first is exemplified by oceanic birds that nest as monogamous pairs in closely packed colonies. In our consideration of fishes, we are not dealing with aggregations of monogamous pairs, although colonial breeding of this sort has been reported for cichlids of the genus *Tilapia*. The second is exemplified by passerines such as weaverbirds and some blackbirds, in which a number of females mate with and nest within the territory of a single male.

Such polygynous systems are closer to lekking as we define it and may even grade into it in fishes. But they suggest more appropriate comparisons with the harem societies of some mammals, such as occur in some pinnipeds and bovids, and among teleosts, of some cichlids of the genera *Lamprologus, Nanochromis, Teleogramma, Apistogramma* and *Nannacara.* Reproductive adaptations based on lekking were first described in birds. Because it is an extreme departure from the usual avian pattern of monogamy and joint care of the brood, lekking has long attracted the attention of behaviourally oriented ornithologists, form the pioneer studies of Selous to recent papers by *Robel* and *Ballard* and *Boag,* and *Pitelka, Holmes,* and *MacLean.* Concise descriptions of what appears to be lekking in teleost fishes were first published by *Reeves* and *Newman,* and have appeared consistently in the ichthyological literature up to the present.

Table 8.1. Incidence of Lekking Among Teleost Fishes

Order: Cypriniformes	Order: Perciforms
Characidae *a*	Centrarchidae
Cyprinidae	Percidae
Catostomidae	Sparidae
Order: Atherinomorpha	Embiotocidae *a*
	Cichlidae
Atherinidae	Pomacentridae
Melanotaenidae	Labridae
Cyprinodontidae	Scaridae
Poeciliidae	Acanthuridae *a*
Order: Gasterosteiformes	Callionymidae
Gasteroteidae	Belontiidae

a Included on the basis of anecdotal accounts describing probable lek systems.

However, no explicit analogy was made between lekking behaviour in fishes and birds until *Fryer* and *Iles* pointed out the correspondence in the reproductive adaptations of many species of the cichild genera *Sarotherodon* (formerly *Tilapia*) and *Haplochromis.* Lekking, or what seems to be lekking, has been described in a variety of other kinds of animals. It has been reported in a number of African antelopes living in open country, and most notable in the Uganda kob, in a variety of other ungulates, in hammerhead bats, in one reptile, in one anuran amphibian and among insects in drosophilid flies, dragonflies. One species of harvester ant also appears to engage in lekking. Although lekking may prove more prevalent among some of these groups than the literature would indicate, fruitful comparisons of lekking in different environments must draw at present upon the ornithological and ichthyological literature. To aid the reader, we enumerate at the outset the prerequisites to lekking any species.

1. *Lekking ground.* The males must congregate at a given place, and the females must also proceed to that place.
2. *Feeding.* Either no feeding occurs on the lek, or only incidental feeding that is insufficient to meet energetic needs.
3. *Synchrony.* The reproductive activities of a substantial proportion of the males and females of a given population must be in phase.
4. *Parental care.* If parental caretaking exists, it requires only one parent, not a pair.
5. *Mobility.* The species must be sufficiently mobile to travel to the mating ground.

Further on we provide some tables that outline the major points inn our central arguments. Although they are redundant to the text, we thought they would be tables, we were motivated to provide testable propositions in the hope that these would stimulate observations to refute them. The reader will discover that the propositions are a pragmatic mixture of inductive conclusions, though often based on scanty evidence, and deductions that seem to us to flow from what we have learned. We now summarise the main features of teleost and avian reproductive biology. Next we compare lekking as practised by members of these two groups. In closing, we speculate on the ecological factors that may have led to the evolution of lek systems in both and attempt to account for its prevalence in teleost fishes.

A Comparison of Teleost and Avian Reproductive Biology

Teleosts are characterised by a wide range of reproductive modalities. Oviparity with external fertilization, ovoviviparity, and true viviparity have been reported for the group. With the exception of the viviparous surfperches (Embiotocidae), for which we have only circumstantial evidence of lekking, and a single viviparous poeciliid, *Poeciliopsis occidentalis*, all lek fishes known to us are oviparous. We will therefore concentrate on the epertinent features of oviparity with external fertilization. In most cases, the female performs a number of spawning acts, releasing some portion of the total spawn each time.

In a lekking species this means that the female may move from one male to another, leaving some eggs with each one. Alternately, she may perform all the spawning acts with one male. The numerous eggs are large and full of yolk. After spawning, the female may not spawn again for several weeks, or until the next season, because it takes time to obtain enough food to lay down such a generous energetic larder. Alternatively, as appears to be the case in cyprinodont, melanotaeniid, and some atherine fishes, the female lays just a few eggs each day over a protracted period. Males, in contrast, need to commit only a small amount of energy to produce vast numbers of tiny sperm. They can spawn repeatedly during a day and for several days. Females therefore make a large initial investment per gamete, the males a small one. A male may fertilize the eggs of a number of females, regardless of whether the spawning pattern involves production of demersal or pelagic eggs,. Thus a male often accumulates a clutch that may number up into hundreds or thousands of eggs, if he is the custodian. Such a number is huge compared to the clutches of birds. The eggs of teleost fishes are surrounded by a permeable membrane of variable strength.

Substances necessary for the development of the embryo, such as water and oxygen, cross this membrane from the external environment while metabolites move in the opposite direction. Teleost eggs are susceptible to attack by bacteria and fungi because their dependence on a steady exchange with their surroundings has in most instances precluded the evolution of effective morphological barriers against such infections. Reproductive success among those teleosts that put their eggs on the substratum therefore depends critically on the accessibility of hygienic spawning sites. Typically, such sites are characterised by a distributed or inherently depauperate microbial community as well as adequate dissolved oxygen to allow normal embryonic development.

Many fishes exploit recurrent natural phenomena, such as the scouring effect of the spates resulting from the spring melt in the temperate zone, or the monsoons in the tropics, to deposit demersal eggs upon the favourable substrata thus produced. Most nest-building species themselves create a substratum with a disturbed microbial ecology through their nest-building activities. A different set of adaptations accompanies the releasing of pelagic eggs into the plankton. In reef-dwelling species that do so, it becomes important to expel and fertilize the eggs at the best place to assure that the zygotes are swept into the most favourable water mass. This often means spawning at the outer edge of the reef when the tides and currents are propitious. Teleosts are poikilothermic. The normal development of their eggs therefore depends on the ambient temperature. This imposes a marked seasonality on the reproductive cycle of temperature zones fishes.

Seasonal changes in water temperature does not seem to play an important role in regulating the spawning of many tropical fishes. Instead the periodicity of rainfall appears to impose seasonality on most tropical freshwater fishes, for whom the arrival of the rainy season often provided the proximal stimulus for spawning. Seasonality seems less pronounced in tropical marine habitats, but the intraseasonal cyclicity of spawning of some fishes in such habitats may be linked to lunar cycles. Taken together, these environmental factors often impose a degree of synchrony of sexual activity, from modest to great, even among coral reef fishes with prolonged breeding seasons. Many fishes, especially freshwater species, practise parental care of their spawn. The caretaker is almost always the male. When the female is involved, the course of evolution appears to have gone from exclusively male care to joint care, then to exclusively female care. That the male is the usual caretaker in fishes stands in contrast to the situation within the Vertebrata, with the exception of the Amphibia.

A further contrast id that there is no reversal of sex roles when the male fish is the caretaker, contrary to *Wilson*. The courtship behaviour of the male fish is masculine by the accepted standards. A key factor in the evolution of such a pattern is that in species that fertilize the eggs externally, the male can be certain that the zygotes left in his care were fertilized by him. Chances of cuckolding are slight compared to species that fertilize internally, the usual case in most of the other vertebrata and in some fish groups. Thus an externally fertilized male can significantly promote his genetic investment by protecting his offspring. *Dawkins* and *Carlisle* have proposed a different explanation for why, among fishes, the male is the usual custodian. Their thinking also turns on external fertilization and is therefore applicable as well to aquatic amphibians with parental care.

When fertilization is internal, "After copulation, the female is left physically in possession of the zygote, and while it is still in her body, she cannot desert it, but the male can. However fast she lays it, the male is still offered the first opportunity to desert, thereby closing the female's options and forcing her into Triver's cruel bind." The bind, as modified by *Dawkins* and *Carlisle*, is that the partner that deserts first does not necessarily condemn the progeny to death. Instead, it simply sloughs the decision off onto the partner left with the zygotes. In fishes and many amphibians that fertilize externally the sperm and lighter than the eggs and are hence more readily dissipated or swept away. From this, *Dawkins* and *Carlisle* reasoned that males have more to lose by spawning too quickly, on the chance that the partner delays, than do females. Thus a female can afford to go first in spawning, leaving the male with the zygotes after he has fertilized them.

In our collective experience, though synchronous ejaculation of gametes seems the general case, whenever one sex spawns first, as in gobies, blennies, and damselfishes, but not in cichlids, there is usually paternal care of the spawn. This explanation, nevertheless, has some problems in wide application. First, one would expect males to evolve sperm that would not wash away, comparable to the spermatophores of some salamanders. In fact, this solution appears to have been evolved by two maternal mouthbrooding cichlids, *Sarotherodon macrochir* and the Tanganyikan endemic *Opthalmochromis ventrailis*, although it does not appear to be inflexibly linked to prior ejaculation by the male. Second, in many maternal mouthbrooding cichlids, the female takes the eggs into her mouth before the male fertilizes them, immediately after she has expelled them. Fertilization occurs intrabuccally.

This suggests that she is eager to possess the eggs, probably in order to minimise the time they are exposed to egg predators, not that she is eager to desert them. Recall, too, that maternal mouthbrooding has probably evolved from joint male-female parental care of the spawn. Finally, it overlooks the general case in seashores and pipefishes, in which the female *inseminates* the male by leaving her eggs in his brood pouch. The explanation proposed by Dawkings and Carlisle provides part of the answer to the prevalence of paternal care of the spawn in fishes. But we suspect that the necessity of sequetering a suitable spawning site (prerequisite to reproductive success in any species with demersal eggs), the energetic differences between the sexes in the production of gametes, and the certainty of paternity that follows external fertilization, are of equal or greater importance in its evolution.

We shall return to this point. Yet another general feature of teleost reproduction needs mentioning. When the male is the exclusive guardian of the zygotes, the female is driven away from the breeding site as soon as the spawning act has been accomplished. Females are notorious egg predators, as seen in cyprinodont fist and gouramies.

Excluding all females save those immediately ready to spawn, thus increase the survivorship of the male's offspring. In addition, spawned-out females might interfere with subsequent mating by the male with other females. Chasing away the spent female is so universal among teleosts that it is remarkable that some groups have been able to evolve joint parental care of the spawn. Parental care in fishes appears in most instances to have been derived directly from the defence of a territory for reproductive or other purposes. It provided a suitable environment for the development of the zygotes, for their defence, and less commonly, for defence of the mobile fry from predators. The eggs and newly hatched fry are fanned and/or mouthed, while predators are driven away by the guardian's attacks.

In most teleosts the period of parental care is brief, extending only to the eggs and larvae. In the few instances where parental care of the mobile fry is practised, a behaviour whose occurrence is limited almost entirely to fresh water fishes, such a commitment rarely lasts more than six weeks. Defence of the eggs or young differs from that seen in birds and mammals. By comparison, the eggs and fry of fishes are tiny relative to the size of the adult, and number at times into the thousands. Their predators are therefore relatively small, seldom larger than the parent and usually much smaller.

Consequently, the parent can readily drive away individual spawn predators at little risk to himself, although his defence may sometimes be overwhelmed by sheer numbers of them. Adult fish are themselves subject to a different type of much larger predator. Thus, when considering the effects of predation upon teleost reproductive patterns, one must keep in mind that it operates upon two different levels, the spawn and the parent, and in ways that elicit radically different responses by the breeding adults. An additional important factor is that much predation on the eggs, and particularly upon the fry, may be by conspecifics. This is true of cichlids and of other freshwater fishes such as sticklebacks and pupfish. Such intense predation upon the eggs and young by conspecifics is less prevalent in birds and mammals.

As a consequence, there is sometimes a semantic difficulty in talking about territorial defence because driving away conspecific intruders may actually represent defence of the spawn against such predators. Although there are too few data available at this time to permit detailed comparisons, it is worth mentioning that territorial defence in lekking fishes may be more ritualized than in non-lekking species. *Apfelback* and *Leong* compared reproductive aggressive behaviour in three species of *Tilapia.* These were *Tilapia zilli*, a monogamous substrate breeding species; *T. galilaea* (=*Sarotherodon galilaeus*), a pair-forming species in which both sexes engage in mouthbrooding; and *T. macrochir* (=*S. macrochir*), a lekking species with maternal mouthbrooding. Aggression was the most damaging in *zilli* and the most ritualized in the lekking species *macrochir*; *galilaea* was intermediate.

Future studies should be alert to the possibility that a concomitant of the evolution of lekking is a shift from damaging aggressive behaviour to ritualized threats. There is no trophic component in the parental behaviour of the majority of teleosts practising defence of their spawn. In some species, e.g., substratum-spawning members of the family Cichlidae, the parents may protect the young as they forage. In cichlids of the genera *Symphysodon* and *Etroplus*, however, and in two species of bagrid catfishes, the fry depend on parental mucus for some of their nourishment. Nonobligatory feeding on parental mucus occurs in a number of other cichlid species. In no instances known to us does parental care include a thermoregulatory component. All birds are oviparous, with internal fertilization of a cleidocal (shelled and self-sufficient) egg. The shell of the avian egg encloses a milieu in which all the raw materials of embryonic development are present save oxygen, and within which provision is made for isolating nongaseous metabolites from the developing embryo. The shell is permeable only to gases and provides a virtually impregnable barrier to bacterial invasion. However, because birds are homeothermic, their eggs require a reasonably constant temperature for development.

Care of the developing eggs, predominantly thermoregulatory in nature, is consequently universal among birds. With the unique exception of the Australasian family Megapodidae, birds accomplish this end by brooding the eggs. During this period, the eggs and one or both parents are vulnerable to predation. Security for the clutch and the incubating parent or parents is consequently a major factor influencing the evolution of avian reproductive adaptations during the incubation period. That period may last as long as the entire interval of parental care in most teleosts. *Lack* has concisely summarised the nesting adaptations adopted by birds to maximise reproductive success. His account underlines the importance of a secure nest sit and various adaptations favouring crypticity for the nest, clutch and parents. Nest site selection is therefore important in birds, but for reasons different than those dictating such behaviour in teleost fishes.

Post-hatching roles vary considerably between nidifugous and nidicolous birds. A thermoregulatory component of variable intensity, nonetheless, is characteristic of virtually all birds. In is most pronounced in nidicolous species, whose young are not feathered at hatching. The precocial young of nidifugous species, in contrast, have a cost of insulating down from the time they hatch. That lessens the need for heat from the parent and is therefore less restrictive of the caretaker's activities. The major difference between these types of birds ,however, is the way the young get food. Nidifugous birds do not usually bring food to their active young, but rather guide them to food sources and provide them with some protection against predators as they forage (*Lack*, 1968). The parents or parent of nidicolous young must forage for food, which is then brought to the young at the nest.

The trophic dependence of the young persists until they are fledged and, in some species, even for a short time thereafter. Regardless of how they are discharged, the trophic responsibilities of breeding birds are of central importance in their overall reproductive strategy. The seasonality so clearly evident in avian reproduction is imposed in large measures by the necessity of having sufficient food at hand both to support the parents and to feed the young during a protracted period of growth, considerations that also limit the choice of nesting site.

Parental role of Lekking in Birds and Teleosts

In all lek birds the male has no parental role. To many ornithologists this is a *sine qua non* for lekking. Once the female has mated and completed her clutch, she avoids the lek, incubates her eggs, and rears her hatchlings alone. The ability of the female to discharge all parental functions unaided is therefore a precondition for the evolution of lekking in birds. Among Lekking teleosts, four possibilities exist with regard to paternal care:

Noncustodial lekking

Demersal eggs are deposited by one or more females within the male's territory on the lek with no prior preparation of a nest, as in most cyprinodonts and in the atherine families Melanotaeniidae and Atherinidae. Alternatively, pelagic eggs are shed into the plankton in a *nuptial dash* launched from the reef, as in the lekking species of the marine families Labridae and Acaridae, and possibly surgeonfishes of the genus *Naso.* There is no overt defence of the spawn, either because of the brevity of the lek's persistence, of siting the lek where there are no predators, or because of the dispersal of the eggs and Larvae.

Lekking with Maternal care of the Spawn

This mode of lekking is known to occur only in one poeciliid and in the maternal mouthbrooding species of the family *Cichlidae.* In the poeciliid, the female visits the male and is inseminated on her court. In the cichlids, the female visits the male and is inseminated on his court. In the cichlids, the female visits the male on his territory, where spawning occurs. The eggs are taken into the female's mouth, wherein they are fertilized if they have not already been fertilized. The egg-laded female leaves the lek while the male remains, awaiting further mates. The female swims to spatially separate nursery grounds, where she remains until the fry becomes independent. Or both sexes may abandon the lek and reconstitute the original school, as in several Lake Tanganyika open water cichlid species. In some instances the fry are shepherded and protected by the female foe a while after they have emerged from her mouth. In others, the young are simply released and abandoned by the female.

Protocustodial lekking

The male constructs a nest that is the focal point in his territory. Then females approach individually and deposit their eggs there. The spawn, of one of several females, are neither cleaned or aerated by the male. Whatever protection from predation the eggs and larvae receive derives incidentally from the male's defence of his territory from conspecific intruders. The fry depart as soon as they are mobile and thus are not protected. In the example known to us, the males abandon their territories after completing the process of reproducing. This mode of lekking occurs in many cyprinids, in the pupfishes of the genus *Cyprinodon* (Loiselle, in prep.), and in some darter perches of the subfamily *Etheostomatinae.* It may also occur in the cod *Gadus callarias.*

Paternal custodial lekking

The eggs of one or more females are deposited within the male's territory in a nest prepared for that purpose. The eggs are usually aerated and cleaned by the male, who vigorously repels all intruders. Some defence o the mobile fry is possibly. This mode of lekking is practised by some cyprinids, by one cyprinodont, Jordanella floridae, by many darter perches, and by all lekking centrarchids. There is some question, however, about the male parental role of one centrarchid fish, the sacramento perch, *Archoplites interruptus.* Further examples of paternal custodial lekking are provided by sticklebacks by some lekking wrasses that produce demersal eggs, and by lekking species of the marine families *Sparidae* and Pomacentridae. In Thailand, groups of male gouramies of the genus *Trichogaster* form discrete arenas under their bubble nests where they are visited by females ready to spawn.

The thermoregulatory and trophic components of avian parental care preclude the evolution of noncustodial, protocustodial, and strictly paternal custodial lekking in birds. These two considerations, taken with the practice of internal fertilization, produce the sharpest difference between lekking birds and fishes. They contribute importantly to the greater diversity in forma of lekking in teleost fishes. That so few internally fertilized fishes have been unequivocally reported to lek may be an accident of insufficient observation or of observations unguided by hypotheses about reproductive strategies. We expect additional examples of lekking to be found among internally fertilizing fishes, as we have suggested for the Embiotocidae. An obvious group to examined the Goodeidae. They are in the same suborder as the Cyprinodontidae, which has so many examples of lekking species; the two families also resemble one another in ecological adaptations and in morphology. We also anticipate the field studies will reveal lek systems to be more prevalent among poeciliids than present evidence would suggest.

Polygamy among Lekking Birds and Teleosts

The absence of monogamous pair-bonding and the corollary occurrence of sequential ploygyny are taken as defining elements of avian lekking. The extent of polygyny is precisely known only in the ruff and in various grouse species. The occurrence of polyandry is also possible in all lek birds but improbable in the galliformes because their females practice sperm storage and hence need visit the lek only once in a season to produce a clutch of the fertile eggs. However, there are no data on polyandry in lekking birds. It seems a probably corollary of the attenuation of the pair bond. Lekking teleosts are likewise characterised by polygamy. Sequential polygyny is found in all lekking species, and sporadic instances of simultaneous polygyny have been reported among the Centrarchidae and the Labridae.

External fertilization of the egg makes this departure from the typical pattern possible. Resident males, however, are normally receptive to but a single female at a time. In some instances, the male may actually repel females that attempt to enter his territory while he is engaged in the terminal phases of courtship or actual spawning. The occurrence of polyandry among lekking teleosts is better documented than is the case among lekking birds. Polyandry has been reported among cyprinodonts and occurs among melanotaenids under quarium conditions (Loiselle, unpublished observation). Females of four darter perches practice polyandry. Such behaviour has been reported as normal in one cichlid, *Sarotherodon macrochir* and in one pomacentrid, *Chromis multilineata* considered it typical of the reproductive behaviour of most centrarchids.

Subsequent investigations have revealed polyandry in one sunfish not cited by Breder, the Sacramento perch. Less information is available on the extent to which individual females of a given species indulge in polyandry. Ruwet reported females *S. macrochir* carrying a clutch of eggs fertilized by five or six males. *Breder* regarded centrarchid breeding systems as essentially nonassortative, citing an instance in which a female *Lepomis gibbosus* visited every male on a small lek of indeterminate size.

Parental Choice, and Predictability of the Environment

The disproportionate reproductive success enjoyed by centrally located males of the ruff and many grouse species with its overtones of Darwinian sexual selection, has attracted the attention of many workers. The occurrence of such a position effect within the lek has not been documented in other lekking birds, however. This lack of information is particularly marked for forest-dwelling lek birds. Until these species have been more extensively studied, it would be premature to regard such position effects as being a universal feature of avian lekking. In those avian species for which position effects have been determined, the classical cases of lekking among tetraonids, central males may enjoy in excess of 80% of the copulations that occur during the breeding season. Succession to central sites within the lek follows a clear protocol. If vacancies occur through mortality, they are filled by peripheral males, who are in turn replaced by marginal males. Direct competition for such sites is ritualized. The protocol of succession resembles the seniority system of the American Congress.

The chairmanships of powerful committees come almost automatically to those who succeed in assuring their regular reelection and avoid antagonising their colleagues by displays of nontraditional behaviour. To the best of our knowledge, however, the analogy breaks down in all but a few cases upon consideration of the rewards accruing to persistent males. There is little information, in studies of fish lekking, on the mechanisms determining access to favoured territories. The evidence suggests that in some centrarchids and cichlids overt competition exists and can be intense. There is no indication of a protocol of succession to favoured sites. There are only fragmentary indications that position effects characterise lekking in teleosts.

The existence of discrete classes of central and peripheral males may be inferred for two lekking cyprinids, one cyprinodont, two darter perches, two centrarchids, gouramies of the genus Trichogaster and one cichlid. *Hunter* reported that leks developed around the first male green sunfish to spawn, and *Wright* observed that male gouramies place their nests around that of the most aggressive male. In Mediterranean wrasses of the genus *Crenilabrus*, a central male is surrounded by smaller nonterritorial satellite males who only occasionally fertilize eggs in the nest when the large male is temporarily away; however, it is not clear whether there is a position effect among the territory-holding males. The extent to which highly dimorphic dominant males enjoy augmented success in mating has been demonstrated well in only one teleost species, the bluehead wrasse (*Thalassoma bifasciatum*), by *Warner* et al. (These males are central in the sense that they are often surrounded by smaller nondimorphic males). The females move across the reef from the shallow to deeper water to each the lek. There the large station-holding males are sought out. As the females approach the lek, they are solicited by the small drab males who are not territorial.

Sometimes females spawn with groups of these smaller males. Still other drab males, *streakers*, join the female when she spawns with the gaudy *central* male, and yet others, *sneakers*, try to steal a spawn on the lek. Nonetheless, the central dominant male enjoys as enormous reproductive advantage over the small drab ones, regularly spawning about 40 and occasionally 100 times/day. In contrast, the nondimorphic peripheral males, spawning predominantly in groups, only achieve the equivalent of about one to two pair-spawnings/day. The situation is less complex in gouramies of the genus *Trichogaster*. Among Thai populations of *Trichogaster trichopterus*, up to 20 to 30 males nest together. The nests are more closely placed around the central male, and each nest territory there is only about 20 cm in diameter. The female swims directly to and butts the male of her choice. In all 88 spawnings observed by *Wright*, the central male was chosen by the female. The example of the bluehead wrasse draws attention to another important difference in teleost lekking made possible by external fertilization.

It is the possibility of neighbouring resident males, or nonterritorial marginal males, joining the consorting couple at the moment of oviposition and participating in the fertilization of eggs. Such behaviour has been documented in one cyprinid, one darter perch, one centrarchid, and one cyprinodontid, one wrasse, and has been observed in one cichlid (loiselle, unpublished data). It is difficult to evaluate the significance of such a breakdown in the lek system on the basis of these examples. One would wish to know, in particular, whether the relative paucity of reports is an accurate reflection of rarity of such a breakdown, or simply an indication of failure to record its occurrence in other species. An instance of a system in which selection seems to have operated against such cheating is cited by Ruwet for *Sarotherodon macrochir*. Resident males whose territories adjoin will display frenetically to an approaching female.

Once the female has entered the territory of one of the competitors, however, all display by the unsuccessful rival ceases. They turn away for the female and indulge either in nest-maintaining behaviour or interact with resident males. In some lekking fishes, such as wrasses and parrotfishes, a form of cheating may be a regular feature of spawning. In some parrotfishes, and in some wrasses, the situation is complicated by a combination of intra- and intersexual competition. Many individuals are sequentially hermaphroditic. Some start life as males (primary males), but most are first females, who then change into males (secondary males). Large males are gaudy, and they lek. The young but sexually mature secondary and primary males resemble the drab females and are the cheaters. They are divergent in the context of this paper, and considering the implications of their biology here would carry us away from the main theme. The existence of position effects among some lek birds and their probable occurrence in fishes raises the issue of female choice. Reproductive age, is the measure of evolutionary fitness common to all organisms.

In lekking and most other promiscuous birds, the only measure of male reproductive success available to the observer is the number of successful copulations per breeding season, since the male plays no role in the rearing of the young. Many of the behavioural and morphological features that characterise males of lek birds, such as extreme sexual dimorphism and elaborate displays, are thus adaptations serving to maximise reproductive success. The measure of female success, on the other hand, is the number of young she brings to independence. This will be determined by her own experience, crypticity, and ability as a mother, and by the genetic endowment of the hatchlings themselves. The latter is the only respect in which the male may make a significant contribution.

It therefore behooves the female to select a male whose genetic material will maximise the chances of her young attaining independence and thus, presumably, sexual maturity. In the grouse the question of choice is yet more crucial, because the sperm storage means a female probably has but one chance a year to make am optimal choice or a mistake. The exigencies of female choice and the nature of the mechanism that determine male succession to central sites within the lek may explain the existence of a position effect by some lek birds. When male succession is largely a function of age, a central male must possess a genome well adapted to its immediate environment. Otherwise he would not have survived long enough to attain such a rank.

As *Wiley* (1974) has shown in one instance, the displays of older birds appear more attractive to females, thus providing a proximal behavioural mechanism for their selection of central males as mates. Females may choose between different groups of communally displaying males. Females of the ruff preferentially visit arenas with a large number of satellite males. As the number of satellites present on an arena declines, so do the number of female visits and the number of copulations enjoyed by resident males.

Additional evidence comes from a colonially nesting species, the village weaverbird. Colonies with fewer than 10 displaying males attract disproportionately fewer females than do the typically larger colonies. Leks may be more attractive to females, and therefore to other males, in direct proportion to the number of displaying males, a point to which we will return. Position effects, as indicated, are not as well documented in teleosts. One would predict their existence in the following situations:

Species practicing protocustodial or paternal-custodial lekking

When defence of the spawn is practised in conjunction with lekking, the optimal strategy for a female is to mate with a male who can provide the most protection at the best location. In a lek situation, central males incidentally benefit from the screening provided through the interaction of peripheral resident males with intruders. Because they have fewer potential predators to contend with, central residents can render more effective defence of their spawn from the few intruders that penetrate the territories of peripheral or of satellite males. At the same time, they can devote a proportionately larger amount of time and energy to actual courtship, thus providing a proximal mechanism of female choice.

Lekking in relation to predictability of the environment

In this section, and in later ones dealing with tradition and evolution, a key concept is predictability of environment. The present digression is necessary to explain how the term is employed. We use the concept relatively loosely and at times as being synonymous with stability of environment. For a more precise treatment of the concept as applied to periodic phenomena, the reader is referred to *Colwell* (1974). Predictability with regard to lekking sites means merely that each year during the breeding season the same set of conditions is apt to prevail on the same lekking grounds. Thus in many species of grouse, the cocks are able to use precisely the same arena year after year. Contrast this with the situation among gouramies of the genus *Trichogaster* in

Thailand. These fishes commence breeding with the onset of the rainy season.

The males build their bubble nests among floating and emergent vegetation in canals, pools, and flooded fields. As the rain continues, the rising water level submerges the vegetation at the original lek. The fish then decamp to find new, better suited sites. Thus, the best place to lek is unpredictable. The seasonal pattern of rainfall, and consequent general pattern of movement of the fishes, however, are fairly predictable. Thus, while the environment may be too unpredictable for traditionality of lek sites to develop, it may be highly predictable in the sense of a given male having the best genetic endowment for coping with it. This brings us to the issue of predictability as a factor influencing female choice. Consider the female's problem: If the environment id relatively predictable, she should mate with the male best adapted to that situation.

The predictability of the environment implies that the genome that is now the most fit will continue to be so in the next generation. If, however, the females' offspring are likely to find themselves in an environment or environments that differ from the present one, the female should not invest all her gametes in the male best adapted to the present situation. Her optimum course of action is to mate with a number of different males. By thus increasing the genetic variability of her offspring, she increases the probability that some of them will be optimally endowed for whatever environment they find themselves in. Predictability of environment is relative to the species, as is the concept of the niche. Take the case of planktonic larvae of a marine fish that are widely dispersed to coral reefs scattered about the tropical sea. A small sedentary species, such as any of several damselfishes or gobies, faces a highly unpredictable community of other species of sedentary fishes when it settles out of the plankton onto a small coral head.

In contrast, a large species such as a surgeonfish of the genus *Naso* can move about and average out local differences in community structure. Its environment is more predictable. Similar arguments can be made for temporal predictability. If the climate is characterised by long cycles of suitable weather and water conditions, the female should pick the currently best adapted male, all else being equal. But if the onset and length of the breeding season and other features of the environment affecting the survival to maturity of the young are unpredictable, the female should be relatively polyandrous. Annual cyprinodont fishes illustrate this point well. These fishes are able to survive in ephemeral pools by virtue of their drought-resistant eggs. The eggs are buried in the substratum of the pool; they survive the dry season and hatch with the onset of the next rainy season.

Not all the eggs spawned in a given year hatch with the coming of the first rains of the next. In a proportion of each spawn, the diapause, or resting stage, of the embryo is prolonged ; from several weeks up to, in some instances, several years. This is an adaptation to environments where the onset of the rainy season is characterised by one or more false starts. Even if many eggs do hatch after a light or unseasonal rainfall, and are subsequently lost, some resting eggs will survive the disappearance of the pool and will hatch with the true onset of the rainy season. In a relatively predictable environment, selection will favour females who produce a large number of nonresting eggs. Such eggs will hatch immediately, giving the fry first access to the food resources of their environment. Such early fry can be expected to reach sexual maturity more rapidly than fry hatched later in the season and to enjoy a longer period of reproductive activity, thus producing more eggs.

In an environment where the onset of the rainy season is unpredictable, the reverse should be true. Females who produce a large number of resting eggs will enjoy disproportionate genetic representation in subsequent generation. While we have no evidence that the earliest hatching males or the largest males in a population produce more nonresting eggs than do smaller, later-hatched males, it is reasonable to assume some correlation between these characteristics. There may well be aspects of the male's behaviour that make them variously adapted to competing, depending on whether they enter the population early, mid, or late in the rainy season.

In any event, assuming that males differ in this regard, we would still predict that the optimal strategy for a female attempting to hedge her bets would entail spawning with a large number of males rather than with one or a few individual males. In predictable environments, mechanisms determining male position on the lek may arise that reflect in a direst manner the adaptive value of a particular genome. As an example, if male position is determined by aggressive interactions, older, larger males, and/or those with a superior energy balance, would be expected to dominate in such encounters. Such males could then secure the nest sites that are best for the development of the eggs, and thus attract the most females. Evidence suggesting this situation in the darter perch *Etheostoma nigrum* was presented by *Winn.* Other males should crowd around this most attractive male to maximise their own chances of attracting females, as in the green sunfish, the bluehead wrasse, or gouramies of the genus *Trichogaster.*

The male attributes contributing to such victories are also correlated with adaptation that particular environment: adaptations leading to increased trophic efficiency results in larger size and / or superior energy balance. In a relatively predictable environment, the genomes of such successful males should converge upon an optimal configuration. Females would then maximise their reproductive success by spawning with such males. It should follow that sexual dimorphism is reduced in those lekking species that are polyandrous. However, because of the brevity of mating, selection will still favour a high degree of dimorphism to enable rapid unambiguous recognition of the opposite sex. This can be based purely on colouration and shape (or possibly on sounds or chemicals), with size dimorphism being more important when polyandry does not occur. Thus, we predict that when position effects are marked, the males will be dimorphic for size and for colour and shape.

On the contrary, when females are less discriminating, the males should be dimorphic for colour and shape but not necessarily for size. Size dimorphism should still be expressed to some degree, however, because size can still be important in obtaining a position of the lek. In fact, when the breeding sites are in especially short supply, as may occur where some annual killifishes breed, size dimorphism should be pronounced. We have written as if environments were either clearly predictable or not, which is not the case. There is a continuum of situations between highly predictable and unpredictable environments. Most will be relatively predictable or unpredictable to varying degrees and with regard to different properties of the environment. Consequently, most species of lekking fishes should reflect a mixed strategy. The more predictable the environment, the more prevalent should be dimorphism, polygyny, and position effect, and vice versa.

Persistence of the Lek

Avian leks are typically occupied for part of each day during the breeding season. The resident males spend the remainder of the day foraging. Individual ,ales of the ruff and several lekking grouse are faithful to a particular site within the lek, as are long-tailed and white-bearded manakins. There are no data available on site attachment in other lekking species. In some lekking grouse, males will revisit the lek site during the fall, well outside of the breeding season. Species inhabiting open habitats active on the lek during the early hours of the day.

One such species, the great snipe, even displays at night during the full moon. This is apparently an adaptation to minimise aerial predation faced by birds displaying in the open. The situation is less clear in forest dwelling species. The impression conveyed in the literature is that males of these species are active on the lek during the latter part of the day. In males of the long-tailed and white-bearded manakin, peaks of activity occur at different times in different parts of their long breeding season (*Mercedes Foster*, in prep.) Such intermittent lekking, in which the lek is occupied for only part of each day, is also practised by some teleosts. It is characteristic of lekking cyprinodonts, melanotaenids, and atherinids, and it may possibly occur in the Sacramento perch. The period of sexual activity, as would be predicted in poikilothermous organisms, is correlated with water temperature and usually occurs in the late morning and early afternoon.

Occasionally, however, high temperatures interfere with lekking during the afternoon. In most marine situations, or in large lakes, the temperatures are more stable and consequently less important as phasic triggers, though thermal effects have been reported. Intermittent lekking is feasible in noncustodial species that produce demersal eggs, such as some cyprinodontids, and that habitually breed in environments into which few or no spawn predators penetrate. This allows males to practice intermittent lekking while accumulating eggs in their territories. It is also feasible for species that shed pelagic eggs into the plankton, such as parrotfishes and wrasses. In neither case is the male's presence required to protect the spawn, nor is there an energetic investment in nest construction to be defended.

Intermittent lekking may, nevertheless, be synchronous if for some reason the females have preferred times for spawning. In contrast to some lekking birds, there is no recorded instance of sexually inactive male fish visiting the lek site outside of periods of reproductive activity. This doubtless happens, however incidentally, in cyprinodontids confined to small pools and in other fishes. Continuous lekking, in which the males occupy the lek without interruption for foraging, relying upon stored energy reserves to sustain their activity, is known only in teleosts. (The mating system of penguins and albatrosses, and of pinnipeds, though not examples of lek systems, provide parallel cases among birds and mammals. Continuous lekking is predictable, for obvious reasons, in fish species practising some type of defence of the spawn, be it constructing an elaborate nest or overtly repulsing predators. Sustained lekking in males of maternal mouthbrooding cichlids, however, cannot be thus explained.

It may instead be correlated with the more overt competition for territories within their leks. *Coe*, for example, reported that male *Sarotherodon grahami* that left their territories to forage lost them immediately to other males and had to contest their possession, often unsuccessfully, with the new proprietors. There are few data on how long an individual male retains a site on the lek. *Reighard* stated that sexually active male logperch, *Percina caproides*, spend 10 to 14 days on the lek, then retire to deeper water. *Neil* found that sexually active male *Sarotherodon mossambicus* hold a nest site in aquarium from three to ten days, with a mode around five to six days. Similarly, successful males of the green sunfish have a period of occupancy of around eight to nine days.

In Thailand, each group of lekking male *Trichogaster trichopterus* lasts about one week. Otherwise, it is known only that the males remain on the lek for a substantial period of the reproductive cycle, a period of time that can vary from several hours, as in the Tanganyikan maternal mouth-brooders *Xenotilapia melanogenys* and *X. ochrogenys*, whose spawning is characterised by a remarkable degree of synchrony, to several days at least. As with intermittent lekking, activity appears correlated with water temperature and usually peaks in the late morning and early afternoon.

Environmental Factors as Determinants of Lek Sites

Traditionality of lek areas is one of the most remarkable features of avian lekking. In fact, *Wilson* gives traditionality as a criterion to distinguish lekking from the more general set called communal displaying. *Armstrong* documented traditionality in five galliform species (Argus pheasant, blackcock, prairie chicken, sharp-tailed grouse, and the extinct health hen–a race of the prairie chicken), one charadriiform species (the ruff, and two forest-dwelling passeriform species (greater bird of paradise, Gould's manakin). *Wiley* presented persuasive evidence for traditionality in an additional galliform, the sage grouse, and *Gilliard* for another passeriform, the cock-of-the rock.

Traditionality is also well developed in long-tailed manakins and in white-bearded manakins. In contrast, nontraditionality appears to be an important feature of lekking as practised by one charadriiform species, the buff-breasted sandpiper (Pitelka, personal communication). This dichotomy of traditionality versus nontraditionality may be explained by assuming that the location of avian lek sites is, or has been determined by environmental factors.

In the case of traditionalists, the factors are presumably a predictable environment coupled with the limited number of areas from which effective displays can be presented. These factors are particularly evident in the case of forest-dwelling species. Authors who have reported on the incidence of lekking in manakins, bellbirds, cock-of-the-rock and birds of paradise; emphasize the following : (1) the performance of the displays requires open space; (2) species conditions of lighting are needed to emphasize the distinctive features of plumage; (3) the lek arenas are located in areas where environmental factors have disturbed the continuity of the predominantly closed forest canopy. The situation in marsh- and prairie-dwelling birds may not be as obvious and is open to debate. *Wiley* and *Hogan-Warburg*, however, implied that the traditional lek areas of the sage grouse and the ruff, respectively, were originally positioned in relatively open patches of habitat where edaphic or other environmental factors had thinned out the prevailing assemblage of forbs and grasses. Students of avian lekking regularly report the occurrence of behaviour by resident males that intentionally or fortuitously preserves and perhaps enhances the suitability of the arena for lekking.

Limited lek sites alone, however, could not account for the highly developed traditionality seen in some species. It requires in addition a relatively predictable environment coupled with a reasonably long life span and the ability of the males to remember the location of the lekking grounds. Otherwise, it would be difficult to account for the fact that some sites are used year after year by long-tailed manakins, while other sites that seem to have all the necessary features are not utilised. It would also be difficult to account for the persistence of lek sites when the environments is unfavourably altered.

Table 8.2. Circumstances Associated with Traditionality

A. Physical environment
   1. Relatively predictable
   2. More "best" arenas exist than are generally used
   3. Arenas often modified by the males' behaviour

B. Animals
   1. Dispersed breeding population
   2. Relatively long-lived (at least more than one breeding period)
   3. Delayed sexual maturity in males.
   4. Central nervous system complex enough to allow of learning and memory.

A well-known example was provided by male ruffs who persisted in displaying on old sites that came to lie in a road. In case of the buff-breasted sandpiper, environmental factors impose nontraditionality. The sandpipers' leks are ephemeral and transitory, males gathering and displaying for periods of a week or two in a given spot, then apparently moving elsewhere to repeat the performance. Pitalka (personal communication) suggested to us that the placement of a lek in these sandpipers is influences by the available supply of food for the nesting female and her brood and by year-to-year variations in the physical environment, such as unpredictable patterns of runoff from the melting ice and snow. According to Pitelka's model, transitory lekking permits exploiting a patchy environment by moving over a wide area and settling in to display only where the terrain is suitable and food abundant. This maximises the probability that females impregnated by them will be able to raise their broods successfully in the short Arctic nesting season.

The occurrence of suitable areas depends on a multitude of local climatic factors and is therefore relatively unpredictable. Traditionality would be maladaptive under such conditions. Lekking may also be correlated with population density. *R.R. Warner* and *S.G. Haffman* are testing the following model, which was inspired by observations on wrasses and parrotfishes off the coast of Panama.

A similar model is being developed for other vertebrates by *S.T. Emlen* and *L.W. Oring.* We present the model in abbreviated form, with due apologies. When the population density is low, the prevailing mating system is a territorial harem society. At intermediate densities lekking develops.

At high population densities dominance relationships break down and territories are forsaken a number of females may spawn synchronously each with more than one male in attendance–the *conubium confusum* of *Breder* and *Rosen.* The intermediate population density at which lekking occurs must be considered relative to the species. It is our impression that lekking species of birds and fishes are relatively common. In general terms, lekking is probably favoured by population densities that are relatively high, but not so high that social organisation breaks down. The dependence of many teleosts on a hygienic spawning location for their relatively vulnerable eggs means that environmental factors directly determine the location of lek sites in species practising noncustodial, protocustodial, and paternal-custodial lekking.

Traditionality in the avian sense is not well documented in such species. However, traditionality should be expected among mobile species occurring in relatively stable environments, such as the rocky littoral of the African Great Lakes or the protected coral reef. There the physical factors that dictate optimal spawning sites differ little from one year to the next. (However, the physical features of some coral reefs may at times be drastically altered in regions where violent storms occur. In variable environments, in contrast, lek sites can change in location from year to year. The bluehead wrasse provides an example of a coral reef fish with traditional lek sites. One population of this labrid used the same area as a spawning site over a period of five years. While this represents traditionality in the broadest sense, the reproductive modality of this species introduces complaining factors not encountered on avian leks.

The bluehead wrasse sheds pelagic eggs into the plankton. It thus adjusts the precise location of its spawning site within a general area from day to day, and even within a day, apparently to remain in a down-current zone that favours the fertilized eggs being swept out to sea. Another instance of what may be lek traditionality in the avian sense is provided by the maternal mouthbrooding Tanganyikan chiclid *Cyathopharynx furcifer. Brichard* reported that males of this species construct sand nest on top of flat-crowned rock blocks. Sexually active males may be found using such areas continuously. Though the number of such sites is limited, intraspecific aggression is not pronounced, and resident males visit one another's territories in a manner reminiscent of such lek birds as the ruff and some grouse.

Additional instances of apparent lek site traditionality have been reported for males of several other maternal mouthbrooding cichlids. Substratum-independent or *free-water* spawning behaviour has been described for the maternal mouthbrooding Tanganyikan cichlids *Tropheus moorii*, *Limmochromis microlepidotus*, and *L. leptosoma*. In most instances, however, sexually active males prepare a nest from which courtship is directed and within which spawning occurs. Since their eggs do not remain in the male's nest during their development, these cichlids are not substratum-dependent in the same sense as are teleosts that produce demersal eggs. Even so their emancipation is not complete. Regardless of their pattern of brood care, cichlids, gouramies, and other teleosts are limited in their selection of spawning sites by predation or both the breeding adults and their spawn. Ease of constructing nests will also vary among sites, as will the degree of shelter form wave action. These factors can adversely influence a male's reproductive success by obligning him to expend energy in nest preparation and maintenance that would otherwise be devoid to courting or to prolonged his time on the lek.

The substratum preferences reported for the cichlid fish *Sarotherodon macrochir* by *Ruwet* were probably in response to the suitability of the substrate. And the destruction of *Haplochromis* leks by wave action, reported by *Kirchshofer* and by *Fryer* and *Iles* suggests that shelter is indeed significant in selecting a lekking ground. The place where lekking is done can be important for still other reasons as when mobile reef inhabitants seeks a favourable lauch window through which to shed their pelagic eggs. In some instances, the siting of leks is constructed by adverse physiographic factors. An extreme case is provided by two cichlids. *Sarotherodon grahami* inhabits hot springs in Lake Magadi, *S. alcalicus* the hypersaline Lake Natron, both in the Rift Valley of Kenya and Tanzania. Thermal factors severely limit the area available to sexually active males *S. grahami* while salinity gradients restrict male *S. alcalicus* in their choice of nesting sites.

Evolution of Lekking

Early attempts to determine the functional significance of avian lekking are exemplified by the following passages from Armstrong (1947:225)

> The conclusion is ineluctable that the advantage of arena or lek displays must be very great. It is highly probable that not only is sociality in itself stimulating, but that the psychological effects of pugnacious posturing have a beneficial effect on the race.

These remarks reflect the group-selectionist framework within which many previous workers approached the study of this animal reproductive adaptation and others. In this account, we follow the lead of *Hamilton*, *Maynard Smith*, *Williams* in affirming that the functional basis of any reproductive adaptation is the increased reproductive success of the individual practising it. Accepting this principle, we should e able to explain both the functional significance of lekking and how it evolved. This requires an examination of ecological factors together with an understanding of the limitations imposed by an organisms' reproductive biology.

Avian Lekking in Relation to Feeding Adaptation

While the ability of the female to raise a brood unaided is a precondition for the evolution of lekking, lek birds are not the only species in which the male is divorced from a parental role. Alternative adaptations based on uniparental care of the brood are possible, ranging from harem polygyny to simple *promiscuity*, practised concurrently with or independent of colonial nesting. Such modes of reproduction probably arise when the trophic advantage of biparental care is outweighed by other factors, for example, by the increased risk of predation resulting from the more conspicuous activities of two parents around the next. This applies to lekking species as well. Furthermore, harem polygyny and/or simple promiscuity occur in such groups as grouse, pheasants, and hummingbirds–groups that contain lek species. There must be finer or overlooked differences in environmental factors that correlate with lekking. A brief digression to consider bird territoriality in its best known form will help us bring out a salient difference.

Classical avian territoriality is feasible only if a resource, such as food, is sufficiently concentrated to make its defence economically profitable. In most territorial birds, the male's defence of his territory simultaneously assures the resources adequate to lodge and fledge a clutch, fixes his position in space, and advertises his presence to females. Conversely, the distribution of discrete advertised territories in a spatial mosaic increases the likelihood of a female encountering an unmated, territory-holding male. The system thus maximises the fitness of both individuals in a pair. The situation has to be different in lekking birds. Without exception, they forage widely for dispersed food, ranging from fruit or nectar to seeds and insects. In some instances, not only are the foods dispersed, but they also tend to be unevently distributed in time and/or in space. And the higher the latitude, the more compressed in time is the period when food is maximally available. Such a feeding adaptation has at least four consequences for the behaviour of its practitioners:

Table 8.3. Circumstances Promoting Lekking in Birds

A. Properties of the physical environment
   1. Natural phasic stimuli for synchronisation of mating, e.g., seasonal changes in photoperiod.
   2. Best places for courtship and mating existing apart from feeding grounds, are characterised by:
      (a) Spatial properties that enhances signal propagation
      (b) Lack of ambush sites and/or unobstructed view of approaching predators

B. Trophic considerations
   1. Food resources either continuously or patchily dispersed.
      (a) Promote mobility, lack of site tenacity, and tendency to aggregate.
      (b) Create need for effective long-range communication
   2. Peaks of superabundance
      (a) Remove necessity of biparental provisioning
      (b) Promote synchronous breeding

C. Predation
   1. Promotes clustering of displaying males, which confers some protection upon them.
   2. Promotes either dispersal of camouflaged nests or clustering of nests in a secure place, well away from displacing males.

D. Social factors
   1. Traditionality
   2. Male need satisfied by small space, since territory serves for mating only
   3. Relatively high population densities.

1. The birds must be prepared to move on to a better location when the local supply of food dwindles. They are less likely to evolve the behavioural mechanisms necessary for sustaining large territories, notably high levels of aggressive responsiveness, a large individual distance, and site tenacity.
2. The nature of the food is such that when it is sufficiently abundant for breeding it is superabundant. Defence of the resource is thus economically unrewarding. Further, only one parent is then required, either as the provisioner, as in song birds, or as the caretaker of self-feeding offsprings, as in gallinaceous birds.
3. Regardless of whether the food supply is uniformly dispersed or patchily distributed, there is still the risk of males and females foraging in different areas, or of individuals being widely separated. They need a communication system to bring them together.
4. We assume that there is a best place, or places, for the males to communicate to the females their presence, identity, and readiness to breed. We also assume that males, not females, congregate, because males mate repeatedly, and because the parental sex cannot afford to be conspicuous to the degree demanded by sexual competition. Males will there tend to congregate at, and compete among themselves for, the best sites. Selection will favour the larger, stronger males in such a situation, leading to the evolution of size dimorphis between the sexes. Further, the broadcast range and channel saturation of the communication system will be heightened through the summed activities of displaying males, increasing the individual fitness of each of them.

The initial stages of a trend toward lekking in birds are evident in non-lekking grouse and hummingbirds. There, male possession of discrete all-purpose territories contrasts, with exclusively female brood care. The common denominator of such territorial behaviour is active defence of a suitable display site by the male. Such sites are prerequisite to reproductive success. Birds whose forging patterns precluded the defence of linked feeding and display territories have a serious problem. They must ensure access to essential display sites while maintaining a normal intake of food.

The difficulty is apt to be acute when the physical setting or predation severely limit the number of sites available. Birds other than oceanic species like penguins are evidently unable to store sufficient energy to allow males to occupy their display sites continuously. They have to vacate the sites daily in order to feed. Only two alternative solutions are therefore possible. The first is for the male birds to engage in physical competition for display sites after each foraging trip. Such activity would be bioenergetically wasteful. It would also increase the risk of injury to the combatants and make them more vulnerable to predators–in short, given them a pyrrhic victory. All these factors would reduce individual reproductive success. The second alternative is to lesson overt competition by increasing the threshold of responses to stimuli eliciting aggression. That would permit males to occupy closely adjoing display sites without continual aggression. This solution also minimises the expenditure of energy while diminishing the risk of injury and predation.

Lekking as practised by such forest species as the bearded bell bird and the cock-of-the-rock appears to illustrate this early grade of lek evolution. Their simple groupings of displaying males appear to lack such concomitants of classical avian leks as classes of males, rigid spatial distribution of display territories within an arena, marked position effects, and clear protocols of site succession. Classical avian lekking, as typified by the ruff, blackcock, and sage grouse, is interpreted by us as aggregations of displaying males whose territories are arranged according to dominance relationships.

Contrary to much widely accepted thought, the dichotomy between territoriality and dominance is not sharp. This type of lekking is characteristic of species inhabiting open country. The relatively undifferentiated topography there results in large aggregations of males, providing enhanced "artificial" land marks for females. In such large aggregations, competition for the best display sites would be increased. Selection should favour individual males who can compete for those vital positions while minimising the risks inherent in combat. Such behaviour is apt to produce distinct dominance relationships. Following these hypothesis, all the classic features of such avian leks may be interpreted as manifestations of a hierarchical social structure. Their existence therefore would be facilitated by long-term association of males outside of the breeding season. Those males would maintain or adjust their dominance relationships, obviating the need for high levels of aggression at the of the next breeding season. The delayed onset of male reproductive activity, another characteristic of birds with uniparental brood care, would also help in this respect: juvenile males are neither well equipped nor strongly motivated to contest for sites in the lek. Lekking among bird implies the physical separation of display and nesting areas.

With no further information, one cannot predict whether such birds would lek or, alternatively, from polygynous colonies, as do village weaver birds, in which display and nesting occur in the same area. The missing element we believe to be the nature of nesting sites in relation to predation.

As Crook pointed out, weaver birds concentrate their nests in the few trees that afford both a large measure of protection from predators and proximity to a rich supply of food. Thus, when the predator-prey relationship favours clustered nests, giving the foregoing ecological situation, a polygynous colony is predicted. But when the best antipredator adaptation is dispersed nests, as in grasslands, a lekking society will evolve. The cock-of-the-rock provides an exception, but one that illustrates the importance of an adequate display arena. The females nest in colonies in rather dark caves with restricted entrances. Visual display within them must be ineffective. The males do not engage females there. Rather, they lek in forest galleries where shafts of sun-light strike their brilliant plumage.

We hypothesize, therefore, that the evolution of avian lekking requires first that food be maximally available for a relatively brief period. This leads to synchronisation of reproductive activity. Second, the food must be superabundant, making its defence as a limiting factor economically unprofitable. This also permits males to confirm their activities to courtship and insemination by facilitating the evolution of exclusively female brood care. Third, the distribution of the resource in space favours the evolution of a system of communication that will bring both sexes together. Perhaps the most critical step in this process occurs when communal display develop at the best available sites for their performance, determined largely by characteristic physical features of such locations. This definition of "best" may include the presence of other males. Vocalisations and conspicuous movements of contrast-rich structures may thus undergo a multiplier effect that further enhances the detectability of each participating male by aiding females in finding the assemblage of displaying birds. The behavioural attributes of classical avian lekking will evolve, once this step has been taken, as selection pressures favour those modifications in behaviour that facilitate the close proximity of sexually active males.

Lastly, predation pressure must favour dispersed nesting. This will result in the movement of inseminated females away from the display ground. The end result is the complete spatial separation of display and nesting ground that is characteristic of lekking birds. We also need to emphasize that the foregoing is a general scheme, and that the factors leading to lekking may have played relatively different roles in different species. In particular, we have slighted the possible importance of predation on the lekking birds.

Lekking may have been, and may still be, crucial for birds that live in open country where little cover is available, but where the males must broadcast to attract females. Then, the males might congregate to reduce predation on themselves, as suggested for schools of fish or any aggregating species. In one mammal that leks in open country, the wildbeest, the males have been shown to be subject to heavy predation.

Lekking in Teleosts

The evolution of lekking in teleosts (Table 8.4) is linked, we believe, to dependence upon suitable spawning sites, by the localised substrate for the deposition of demersal eggs or points assuring a favourable launch window for pelagic eggs. While synchrony of reproductive activity is imposed by environmental factors, as in birds, proximity is determined by the availability of suitable spawning sites. Where these spawning are limited in either occurrence or extensiveness, their sequestration, either totally or in part, by an individual may be a positive adaptation. Males would be in a better position to implement such as adaptation because of the greater energetic demand that the maturation of eggs places on females. As in birds, however, total sequestration of a resource is reasonable only if its defence is economically profitable. Hence, while a male fish might, theoretically, enhance his fitness at the expense of conspecific competitors by monopolizing a spawning site *in toto*, such a strategy would be practical only if the energetic cost to its practitioner were exceeded in some manner by a positive return on the investment. Here the positive return would be enhanced reproductive success.

Table 8.4. Circumstances Promoting Lekking in Teleosts

A. Properties of the physical environment
  1. Natural phasic stimuli for synchronisation of mating, e.g., seasonal changes in temperature, rainfall, or lunar/tidal cycles
  2. Best places for courtship and spawning exist apart from feeding grounds. They are characterised by:
     (a) Hygienic properties that promote the development of the zygotes after spawning
     (b) Spatial and physical properties that enhance signal propagation
     (c) Relative freedom from predation due to
        (1) Inability of predators to penetrate the arena
        (2) Lack of ambush sites and/or unobstructed view of approaching predators

B. Trophic considerations
  1. Lack of trophic component in brood care facilitates uniparental care of spawn
  2. Ability to store metabolic reserves allows males to hold territories for extended periods of time

C. Predation
  1. Promotes synchrony of breeding, as numbers of eggs and fry procured can "swamp" their predators
  2. Promotes clustering of displaying males
     (a) Summated brood defence
     (b) Protection against predators of adult fish "swamping" or by "selfish herd" phenomenon

D. Social Factors
  1. Traditionality
  2. Founder-male effect
  3. Male needs satisfied by small space since
     (a) Territory used only for spawning or mating
     (b) Brood care strictly custodial
  4. Relatively high population densities

We can find no record in the literature of any teleost whose reproductive pattern includes such behaviour. Rather, the pattern that emerges is sequestration of a site only large enough to facilitate successful spawning. Because the lek site itself is unimportant in the nutrition of the offspring in virtually all lekking teleosts, the defence of extensive territories is unnecessary. Lekking therefore arises automatically in teleost fishes when the size of a discrete spawning territory sequestered by a male is small relative to extent of the available site. As the number of males entering the area and setting up territories increases, the size of each territory will contract to smallest area required by each male in order to reproduce successfully. Under most circumstances, and for most animals, the resulting geometry will converge upon, but seldom achieve, an hexagonal array of territories. We suspect that intermittent lekking is the most primitive manifestation of this reproductive strategy in teleosts.

As males expend energy in obtaining and defending a site, selection will favour the evolution of site tenacity. From the standpoint of minimising the chance of injuries sustained in intraspecific combat alone, this is a more efficient strategy than repeated contests for a suitable spawning site. The result of such selection would be the appearance of persistent lekking in the absence of any sort of parental care, as is seen in some darter perches. The defence of a spawning territory, even if only from conspecifics, confers serendipitously a degree of protection from predation to the eggs deposited therein. It requires little additional behavioural adjustment to broaden such defense to include heterospecific predators on the spawn.

Natural selection would favour those fish who did practice such defence, or whose spawning behaviour in some other manner favoured the survival of their fry. Hence, the widespread occurrence of some type of protective behaviour among teleosts with demersal eggs. Among teleosts that practice parental care, the primitive condition is for the male to establish a territory and for the female to visit him there. The female then has four options: (1) She can remain with the male, an advance condition shown by few kinds of fishes, and join in the care of the spawn. (2) She can assume full care of the eggs herself, while the male obtains further females elsewhere, as in some of the dwarf cichlids. (3) She can pick up the eggs in her mouth and leave his territory. As previously noted, option (3) is confined to one family of fishes; it has clearly been derived from the first option–that is, joint parental care. (4) She can leave the eggs to the care of the male–the most general case. This solution permits the male in his territory.

However, if selection favours active paternal care, such as fanning the eggs, or assisting them to hatch, or dispersing the larvae, then the male will come to accept eggs for only a brief period in order to coordinate his care with the needs of the brood. Consequently, the reproductive system will become increasingly dissimilar to avian lekking. We believe lekking to be prevented among freshwater teleosts because it is compatible with the third and fourth spawning options presented in the preceding paragraph. Indeed, it may actually facilitate their implementation in certain situations. Among maternal mouthbrooding cichlids, the advantages of lekking behaviour must be similar to those that accrue to lek birds. Among custodial lekking fishes, participants in the lek may well benefit from a summed territorial defence, with centrally located males enjoying a reduced burden of defending the spawn because of the activity of peripherally located males. This factor appears to have influenced the evolution of lekking in gouramies of the genus *Trichogaster*.

Regardless of whether they practice parental care or not, lekking teleosts have the potential advantage of swamping spawn predators with the sheer number of eggs and fry produced within a limited area, depending on the number of eggs and fry, and on predators and their capacity to devour eggs and fry. Thus just as predation upon nesting individuals has combined with feeding adaptations to produce lekking in birds, so has feeding biology and spawn predation, interacting with dependence upon hygienic spawning sites, led to the widespread occurrence of this type of mating system among teleosts.

An Overview of Lekking in Fishes

We would like in closing to be able to make a straightforward comparison of lekking in birds and in fishes (Table 8.5). That is not easily done, and for two reasons. First, there are more variations on the lekking theme within teleost fishes that exist among birds. Second no one species of fish offers a pattern of lekking that is completely comparable to that seen in any bird. Yet all the cases previously cited fulfill our minimum prerequisites for lekking. These are synchrony of the breeding population, lekking grounds where males await females, sufficient mobility to move between separate areas for breeding and feeding, and performance of parental care, if it exists, by only one parent.

We do not include as a minimum prerequisite the absence of habitat constraints. *Pitelka* believes that it is necessary to show that there are available but unused alternate sites for lekking. Even when this can be demonstrated, it does not rule out the importance of suitable sites. For example, the long-tailed manakin apparently has a number of requirements for its arena, including a rare vine that is always present. Yet traditionality is highly developed, and some seemingly suitable sites are not used. The issue is whether some attractive but limited features of the environment occur in only a few places, and thus help to concentrate the animals, or whether the animals themselves act as the focal point, given that adequate situations occur in a number of different places. We see these as interactive factors rather than opposing ones, and factors whose operation is effective to varying degrees, depending on the species.

Table 8.5. Characteristics of Highly Evolved Lek Systems in Both Birds and Teleosts

A. General characteristics
   1. No feeding on the lek
   2. The more males present, the more females attracted
   3. Reduced aggression and increased intermale display
   4. Clear dominance relationships
   5. Well-developed sexual dimorphism
   6. "Cheating" by young or subordinate males rare
   7. Succession to central positions determined by a strict protocol

B. Characteristics of central males
   1. Largest and oldest are the most dominant
   2. Experience less interference with mating or spawning
   3. Devote relatively more time courtship, less to status, fights, territoral defence, or, in teleosts, antipredator behaviour
   4. Females select central males
   5. Central males mate or spawn more than do other males.

A close parallel is exemplified by parrotfishes. The males take up positions on the lekking ground for only a few hours during the day, as in grouse. And the situation is complicated by small males who resemble females and compete with the lekking males to fertilize the females' gametes. Another reasonably close parallel is found in the desert pupfish whose reproductive pattern is probable typical of that of most cyprinodontid fishes. The males apparently leave the lek to feed and to sleep, as do classically lekking birds. There are even peripheral males that could be called satellites.

The critical difference from birds, however, is that the demersal eggs are left in the male's territory, although the male provides no overt spawn care. Most freshwater lekking fishes seem to be of the sunfish type. They differ among themselves in the extent to which they provide parental care, from protocustodial to paternal custodial, and therefore remain continuously on the nest site. This mode of teleost lekking diverges most strongly from the avian paradigm in terms of the persistent occupancy of courts by sexually active males and the existence of a paternal custodial role. The last type of lekking in fishes is that of maternal mouthbrooding cichlids. Cichlids parallel birds in that the females have their gametes fertilized at the lek, and their take them to separate brooding grounds to rear and, in many cases, shepherd the offspring.

Unlike birds, the males stay on the lek for periods of a week or so. And unlike other fishes, the nature of the substrate is obviously not critical for the development of the eggs. The particular site, however, can be important for nest-building and courtship, and for the avoidance of predation. Four factors in the biology of teleost fishes stand out as being responsible for the differences from birds. One is that they regularly store metabolic reserves, a practice that allows the males to remain on station without feeding for long periods of time. The second is external fertilization. The third is the sensitivity of their eggs to the medium that surrounds them. The fourth factor is the huge clutch size of teleosts, with a correspondingly lower investment per gamete. Given this diversity within lekking by different species of teleosts, it is difficult and perhaps premature to attempt to explore how trophic adaptations might be of overriding importance in the evolution of lekking in fishes, as we attempt to do for birds.

However, some obvious generalisations can be made about ecological factors, even if they only serve as hypotheses to be disproved. The trophic adaptations of lekking fishes must require a degree of mobility compatible with moving to breeding areas. The breeding areas must have some features that make them more suitable that those in which other activities take place, particularly feeding; these would include favourable physical conditions for the development of eggs and lessened vulnerability to predation on the fish and/or their eggs. Breeding must also be sharply phasic to bring a number of reproductively responsive animals together; phasing may be through trophic and climatic factors such as rainfall, or through tidal cycles linked to lunar periods. The foregoing general features of lekking probably apply to any kind of animal that utilise a lekking mode of reproduction.

9

AGONISTIC BEHAVIOUR

Aggression is a complex behaviour with many functions and it may include predatory behaviour. *The term with a more precise definition is agonistic behaviour which is a system of behaviour patterns that has the common function of adjustment to situations of conflict among conspecifics.* The term agonistic includes all forms (aspects) of conflict, such as *threat, submission, chases* and *physical combat.*

Forms of a Aggressive Behaviour

Aggression is expressed in two ways:

(i) Social use of space.

(ii) Dominance hierarchy.

But we can list the forms of aggression in the following way:

1. *Sexual.* Use of threats and physical punishment, usually by males to obtain and retain mates.
2. *Moralistic.* Enforced conformity to group standards in humans, often based on religion and ideology.
3. *Predatory.* Act of predation, possibly including cannibalism.
4. *Antipredatory.* Defensive attack by prey on predator such as moblsing.
5. *Territorial.* Exclusion of others from some physical space.
6. *Dominance.* Central as a result of previous encounter, or the behaviour of a conspecific.
7. *Parent off-spring.* Disciplinary action of parent against off-spring.
8. *Weaning.* Restriction of access of offspring to milk.

Aggression and Competition

Most agonistic behaviour *involves competition* for some resource.

The competition is of two types: (i) Scamble and (ii) Contest.

1. *Scamble.* To *win* something. For example, if we throw a price of food between a couple of squirrel monkey they will both go after it, with the first one to reach the food and take it away.
2. *Contest to Compromise.* This will bee well understood by the following example:

If a piece of food is thrown between two *Rhesus* monkeys one would probably grivace and face away. The first would then calmly work and take the food.

Types of Aggressive Behaviour

Now lets consider the two main types of aggressive behaviours in details.

Social Use of Space (Territoriality)

The area used habitually by an animal or group, and in which the animal spends most of its time, is the *home range.* The area of heaviest use within the home range is the *core area.* This may include a nest, sleeping trees, water sources or feeding trees. The minimum distance that an animal keeps between itself and other members of the same species is its *individual distance.* An area occupied more or less exclusively by an animal or group and defended by overt aggression or advertisement is a *territory.*

Size and boundaries of territory

Size depends on the size and diet of the animal as well as on the particular resource the animal is defending. Most vertebrates defend the food source. Territorial animals spend much time patrolling the boundaries of their species singing, visiting scent posts and making other displays.

Territorial model

Carpenter and *Macmillian* (1976) demonstrated in their studies that the possibility of constructing a model to predict territorial behaviour. They argue that in order for territorial behaviour to occur E the basic cost of living, plus T the added cost of defending a territory must be less than the yield to the individual if not territorial plus the extra yield gained by the reduced competition of territorial (fraction of b of productivity P). In other words, $E + T < aP + bP$. Not all territories include all the resources that the animal needs. e.g., *Lek*, a mating system involving a peculiar type of territory if the *Lek*. In this case the resource of the organism defends is the place for mating. On the *Lek* the males do title or no feeding, they spend all their time and energy patrolling and boundaries displaying to other males, attempting to attract females into their area. We can see many forms of territories in animals founded for defending the special, nest building, for mating place, play, or feeding etc.

Dominance

The organisms remember each other by odour or previous encounters which beat them previously and which ones they were able to dominate. This establishes *dominance hierarchies* which requires individual recognition and learning based on previous encounters among the individuals.

Dominance hierarchies

The idea of dominance was first given by *Schjelderup. Ebbe* (1922) who worked out the peck in domestic chickens. Dominance hierarchies from most often in arthropods and vertebrates that live in permanent or semipermanent social groups. Hierarchic means the *ranking of individuals* for different objects such as assessment of water, favoured breeding site etc. An animal is dominant if it controls the behaviour of another (*Scott* 1966). In other sense, a prediction is being made about the outcome to future competitive interactions (*Rowtl* 1974). For example, if four or five mice that are strangers to each other are put together, several outcomes are possible.

Most likely a despot will take over and all the subordinates will be more or less equal. Another possibility is a *linear hierarchy* where A dominates B, B dominates C and so on (i.e., A→B→C→D). Sometimes triangular relationship is formed such as,

```
      A
    ↗   ↘
  B   ←   C
```

In other species coalitions may affect dominance, such that

```
      A
    ↙   ↘
  B       C
  C   +   B
      ↓
      A
```

Dominance hierarchies seem to be a common result when socially living organism engage in context competition. For example, Dominance hierarchy in *Rhesus monkeys.*

Dominance in females

Position in the hierarchy is determined by the mothers rank (*Sade* 1967; *Maxdew* 1968). Adult females are linearly ranked with sons and daughters generally ranked just below their mothers. Dominance in these families manifests itself sublity but constantly at a food or water source or increases to testing spots with the dominants supplanting subordinates.

Dominance in males

Dominance among males is more complex because males have the natal group shortly after puberty while in the natal group they rank below their mothers. As males move in new groups dominance relationship are quickly formed. Size and previous fighting experiences seen not to influence rank directly as long as they stay in the new group, their rank positively correlated with seniority those in the group the longest rank the highest (*Duckenar* and *Nessey* 1973).

Advantage of dominance

(i) Studies of group living birds and mammals indicate that the dominant animals are well fed and healthy.

(ii) *Christian* and *Davis* (1964) have reviewed data for mammals showing that low ranking individuals, frequent losers in fight have higher levels of adrenal cortical hormones than do dominant. These hormones elevate blood sugar and prepare the animal for fight of flight. The cost is a reduction in antigen–antibody and inflaminatory responses–the bodies defense mechanisms–and a reduction in levels of reproductive hormones.

(iii) *Reproductive success.* Higher rank is often directly correlated with reproductive success, as *Le Boeuf* (1974) noted in his study on elephant seals.

(iv) Many popular treatments have emphasized the importance of dominance and aggression in keeping the species fit, since the survivors of fierice battles are the most vigorous and therefore "improve" the species by passing those traits on.

Factors Affecting Aggression

There are two factors which affects aggression :

(i) Internal factors in aggression
(ii) External factors in aggression

Factors

Internal	*External*
1. Limbic system	1. Learning and experience
2. Hormones	2. Pain and frustration
3. Genetic control	3. Lenophobia, crowing breeding, feeding

Internal Factors in Aggression

Limbic system

The brain structure involved in emotion are part of the Limbic system. The hypothalamus is involved in defense and escape behaviour in animals as diverse as pigeons, cats, monkeys and opposums. Researchers have found that different brain sites are responsible for different types of aggression. For example, electrical stimulation, in cats, of the ventromedial nucleons of the hypothalamus produces growing, hissing and attacking with claws.

Hormones

Interacting with the limbic system are neurosecretions and hormones. Epinephrine (adrenation and norepinephrine or noradrenaline) are related to physiological arousal. These are other substances, such as dopamine and serotonin, may act as neurotransmitters and may affect aggressiveness. Early work on chicken demonstrated that testosterone made them more aggressive and increased their dominance rank. Reproductive hormones other than testosterone also affect aggression. L.H. increases in some species of birds. Estrogen lowers aggression in some species and raises it in others (*Davis* 1963).

Genetic control

The synthesis of nervous structures, neurosecretions and other compounds is under genetic control, and agonistic behaviour is shaped by natural selection, as it any other behaviour. Artificial selection can lead to significant changes in level of aggression within just a few generations.

External Factors in Aggression

Learning and Experience

Researches generally accepted the fact that some factor outside the animal triggers the response. Previous experience can produce semipermanent changes in the experssion of agonistic behavioural and animals can be conditional to win or lose. *Scott* (1975, 1976) has emphasized the role of experience and learning in the development of aggression. Although he acknowledges the genetic and physiological bases of aggression. He emphasizes the importance of early experience and learning in the development of aggression and suggests that one of the most important of aggression and suggests that one of the most important controls of agonistic behaviour is passive inhibition.

Pain and Frustration

More direct causes of aggression are pain and frustration.

Pain. Nexious stimuli such as wind noises, foot shocks, tail pinches and intense heat, cause different lab species to attack a wide variety of objects from conspecific cage mates to tennis balls. (*Ulrich, Hutchinson* and *Azrin* 1965).

*Frustration. R*esearchers have explored the frustration aggression convection in humans. In some early experiments researchers allowed children to view a room full of toys but denied them access to it, when they finally led them into the room, the children group that was allowed immediate access to the room. Frustration is now considered to be only one of several causes of aggression.

Xenophobia, Crowding, Breeding, Feeding

Xenophobia. This is another strong stimulus for aggression. Xenophobia means presence of a stranger. In most species, introducing a strange animal into an established social group produces the violent reaction of Xenophobia, usually among members of the same see as stranger (*Southwich* et. al 1974).

Crowding per setends to increase interaction rates but does not always increase aggression. The overall group size, presence of strangers, or restricted access to resources has a much more powerful effect on aggression than dos reducing the peace available to an already established social unit.

Breeding. Courtship and cooperation themselves seen to have aggression components as evidenced by the appaling rackel mating cats make which makes the listener wonder whether it is a fight to death or courtship. Some of the aggression at the time of mating stems from the fact that males tend to court females rather disiriminately, often making advances when the females are not sexually receptive.

Feeding. Some evidence has shown that feeding in groups of fishes intragroup agonistic behaviour. *Albrecht* (1966) proposed that predatory and aggressive behaviours, although functionally distinct are motivationally linked such that motivational summation occurs.

Restraint of Aggression

Displays

Displays that communication an animal's intention and mood may inhibit attack be another *Lorenz* (1966), *Scott* (1966) and others have argued that such behaviours are necessary to avoid large scale killing and wounding.

Evolutionary Model

The species preservation argument has become unpopular recently because it implies that natural selection operates at the group or species level. *Tinbergin* (1951) had previously pointed out that retiralized contests would reduce the risk of injury to the aggression and that threats, which take less energy than physical combat are of advantage to individual as well as groups. Whatever, the evolutionary origins, the ritualization of agonistic displays into contests rather than struggles has occurred in practically all species with social behaviour.

Relevance of Humans

Control of aggression in animals may have special relevance for humans. Humans also become violent according to *Lorenz* Because we have few harmless outlets for aggression. *Lornez* says that if aggression in inevitable then we must find harmless ways to vent it. This cathantic approach suggests the importance or ritualized tournaments e.g., sports events in which participants and spectators alike can work out their hostilities.

Social Control and Disorganization

Scott (1975) has developed the idea of the social control of aggression. Observation of agonistic behaviour in eichlid fish has shown that when strange fish of the same species are continually introduced into the group, fighting remains at a high level, if group membership is allowed to stabilize, fighting declines to a relatively low level. One cause of *aggression* seems to be *social disorganization* when a high ranking male or female *Rhesus monkey* dies or leaves the group, aggression within the group may increase.

The War of Attraction, or 'The Waiting Game'

The Problem

Female dungflies, Sactophaga stercoraria, come to fresh cowpats in order to lay their eggs. Swarms of males are waiting for them on and around the cowpats and whenever a female arrives, the first male to encounter her copulates with her and then guards her wile she lays. Females prefer to lay in fresh dung and as the pat gets older, and a crust forms over it (thus making it less suitable for egg laying) fewer females arrive. The males's problem is, what is the optimum time to spend waiting for females at each cowpat? The answer is that the best waiting time for one male depends on what all the other males are doing. For example, if mist males remain for short times then a male who stayed a little longer would have high mating success because he could claim all the late arriving females. If, on the otherhand, most males were staying a long time then it would pay our males to move quickly to a new pat to claim the early arriving females there. We have to analyse, therefore, a competition among the males where the pay-offs for different waiting times are frequency dependent.

The Theory

This kind of game, where winners are decided simply by a contest involving waiting, or displaying, for different lengths of time is known as the war of attrition and the solution is as follows. Imagine two individual contesting for a resource value V. The longer an individual waits, the greater the cost, m. Let animal A select a time x_A and animal B select a time x_B. The costs associated with these times are m_A and m_B. If $x_A > x_B$ then A wins the resource and so we have.

$$\text{Pay-off to } A = V - m_B$$
$$\text{Pay-off to } B = \quad - m_B \rightarrow$$

Note that A only pays the cost of waiting for time x_B because B gave up first. We suppose now that waiting times are heritable and that individual have offspring with identical waiting times to themselves. We suppose too that individual reproduce in proportion to their pays-offs in the game. How would evolution proceed in this game? Obviously B would have done better to wait for longer than x_A so it could have won. For example, a strategy 'wait for 1 min' would be beaten by a strategy 'wait for 1.1 min.' This might suggest that longer and longer waiting time would evolve (1.2 beats 1.1, 1.3 beats 1.2, and so on) but eventually these waiting times would become so long that the costs would exceed the value of the resource.

It is clear that no 'pure' strategy can be an evolutionary stable strategy, or ESS. (*An ESS is a strategy which when adopted by most members of the population cannot be beaten by any other strategy in the game*) If an individual always played the same fixed waiting time, then other competitors could take advantage of this and play a randomly chosen waiting time, in other words be unpredictable. This stable solution to the waiting game could come about in two ways.

1. There could be a polymorphism in the population with every individual playing a fixed waiting time, or pure strategy, but with the frequencies of the strategies following the random distribution. There would then be a mixture of pure strategies, with most individual playing short waiting times and fewer playing longer times.
2. Every individual in the population could play the same variable strategy, or mixed ESS. Every individual would sometimes wait for a long time and sometimes for a short time, and the times would be selected from the appropriate random distribution.

The first prediction from this model is therefore that *waiting times will be variable*, either within or between individuals. The second prediction is that at the *ESS, the pay-offs for the different waiting times will be equal.* Equality of success will come about through frequency dependent selection. Imagine, for example. That individuals with short waiting times are enjoying greatest reproductive success. Their progeny will inherit this successful strategy and the proportion of individuals adopting short stay times will increase in the population. As it does, however, the success of short stay times will decrease because of increased competition.

As more and more male dungflies, for example, play short stay times, there will be more competition among these males for the early arriving females. The increase in proportion of the short stay times in the population will come to a stop when this strategy would decrease in the population. As it did, its reproductive would increase, because of reduced competition and frequency dependent selection would again stabilize the proportion at that which gave equal reproductive success to the other waiting times. The solution to the game in the equation above is *stable* in the sense that once the population adopts this strategy then no individual who played a different strategy could gain higher reproductive success. The model's assumption of a sexual reproduction ('like gives rise to like') is not too restrictive. Provided the ESS phenotype can be produced by a genetic homogygote it will be univadeable (stable) in a sexually reproducing species with diploid inheritance just as in an asexual one.

Testing the Theory

What do dungflies do? Parker measured the lengths of time that males spent waiting for females on cowpats and the results showed that, as predicated by the war of attrition model, the *stay times* fitted a negative exponential (random) distribution. It is not yet known whether individuals play a mixed strategy. Parker also measured the reproductive success of males and showed that the distribution of male stay times was such that different waiting times resulted in similar mating success, again as predicted by the model.

The Evolution of Conventional Fighting

The war of attrition model is likely to be most applicable to situations, like the dungflies, where individuals are engaged in exploitation competition or '*playing the field*' (*Maynard smith* 1982). Many animal conflicts, however are examples of contest to be solved simply by a waiting game! Indeed, in cases like this there tends to be direct confrontation in the form of fights. For example, after a male dungfly has paired with a female other males may approach and attempt to take her over. There is often a struggle before one eventually wins and the other gives way. Although fights like these can sometimes be vicious in animals, in general disputes are often settled before there is serious injury. Why do animals often settle contests by display rather than by all out fighting? The long accepted answer to this question was that excalated contests would result in many animals getting seriously injured and this would. This is a group selection argument and does not explain how natural selection acting on individuals can give rise to the evolution of conventional fighting.

Since the mid 1960s many people realized that the answer must be framed in terms of costs and benefits of fights to individuals. Once again game theory models developed by *Maynard Smith* and others are useful for exploring the factors influencing the evolution of pair-wise contests.

Hawks and Doves

Consider a game where there are just two strategies 'Hawks' always fight to injure and kill their opponents, though in the process they will risk injury themselves. 'Doves' simply display and never engage in serious fights. Although simple, these two strategies are chosen to represent the two possible extremes we may see in nature. In this evolutionary game, let the winner of a contest score +50, and the loser 0. The cost of a serious injury is–100 and the cost of wasting time in a display is–10. These pay-offs are some measure of fitness and we will assume, for simplicity that Hawk and Dove reproduce their own kind faithfully in proportion to their pays-offs. (The exact values do not matter and are chosen simply because this game is easier to explain with numbers rather than algebra.) The next step is to draw up a two by two matrix with the average pay-offs for the four possible types of encounter. How would evolution proceed in this particular game? Consider what would happen if all individuals in the population are Doves.

Every contest is between a Dove and another Dove and the pay-off is on average +15. In this population, any mutant Hawk would do very well and the Hawk strategy would soon spread because when a Hawk meets a Dove it gets +50. It is clear that Dove is not an evolutionarily stable strategy, or ESS. However, Hawk would not spread to take over the entire population. In a population of all Hawks the average pay-off is –25 and any mutant Dove would do better because when a Dove meets a Hawk it gets 0 (which is not very good, but still better than –25!). The Dove strategy would spread if the population consisted mainly of Hawks. Therefore, hawk is not an ESS either.

Table 9.1 The game Between Hawk and Dove

(a) Pay-off winner + 50 Injury –100
Loser 0 display –10

(b) Pay-off Matrix average pay-offs in a fight to the attacker.

Attacker	Opponent	
	Hawk	Dove
Hawk	(a) $^1/_2(50) + {}^1/_2(-100)$ = –25	(b) +50
Dove	(c) 0	(d) $^1/_2(50-10) + {}^1/_2(-10)$ = + 15

Notes:

(a) When a Hawk meets a Hawk we assume that on half of the occasions it wins and on half the occasions it suffers injury.

(b) Hawks always beat Doves.

(c) Doves always immediately retreat against Hawks.

(d) When a Dove meets a Dove we assumes that there is always a display and it wins on half of the occasions.

Nevertheless, a mixture of Hawks and Doves could be stable. The stable equilibrium will be when the average pay-offs for a Hawk are equal to the average pay-off for a Dove. If the population moved away from this equilibrium then either Dove or Hawk would be doing better and so the population would not be stable. Each strategy does best when it is relatively rare and the tendency in this evolutionary game will be for frequency dependent selection to drive the frequencies of Hawk and Dove in the population so that they each enjoy the same success. For the values in Table 10.1 the stable mixture can be calculated as follows. Let h be the proportion of Hawks in the population. Therefore, the proportion of Doves must be $(1-h)$. The average pay-off for a Hawk is the pay-off for each type of fight multiplied by the probability of meeting each type of contestant. Therefore.

$$\overline{H} = -25h + 50\,(1-h).$$

Similarly, for Dove the average pay-off will be

$$\overline{D} = oh + 15\,(1-h).$$

At the stable equilibrium (the ESS) $\overline{H}$ is equal to $\overline{D}$. Solving the two equations above by setting $\overline{H} = \overline{D}$ gives $h = 7/12$, and therefore, by subtraction, the proportion of Doves (1–h) must be 5/12. Just as in the war of attrition, the ESS could be achieved in two distinct ways:

1. The population could consist of individual who played pure strategies. Each individual would either be Hawk or Dove, and the ESS would come about with 7/12 of the population being Hawks and 5/12 doves.
2. The population could consists of individuals who all adopted a mixed strategy, laying Hawks with probability 7/12 and Dove with probability 5/12, choosing at random which strategy in the game. If the population consisted of a different mixture of individuals from that in (1) or of individuals playing their

strategies with different probabilities from those in (2), then there would be no equilibrium: either Hawk or Dove would enjoy a temporary increase in success until the population evolved to the ESS once more.

It is instructive to note that at the ESS, the average pay-off which individual would enjoy if they all agreed to fight as Doves, namely 15! This makes clear the point we made in the opening paragraph to this chapter. The optimal strategy to maximize everyone's fitness would be an agrement among all to play Dove. However, this would not be stable because in a population of all Doves. Hawk mutants would invade. We expect evolution to lead to stable strategies because, in the words of Richard Dawkins. They are immune to treachery from within.'

Hawk, Dove and Bourgeois

Now imagine another strategy in this game, 'Bourgeois.' With this strategy the individual plays 'Hawk, if owner' and 'Dove, in intruder.' In other words, it fights hard if it's the owner but always retreats if its, the intruder. Let us keep the same pay-offs as before and, for simplicity, imagine that a Bourgeois individual finds itself owner half the time and intruder half the time. With three strategies in the game, the pay-offs are indicated in Table 9.2

Table 9.2. The Hawk-Dove-Bourgeois Game

(a) Pay-offs (as in table 5.1)

Winner	+50	Injury	–100
Loser	0	Display	–10

(b) Pay-off Matrix average pay-offs in a flight to the attacker.

		Opponent	
Attacker	Hawk	Dove	Bourgeois
Hawk	-25	+50	+12.5
Dove	0	+15	+75
Bourgeois	–12.5	+32.5	25

Notes:

1. The top left cells are exactly the same as in Table 9.1.
2. When Bourgeois meets either Hawk, and intruder half the time and therefore plays Dove. Its pay-offs are therefore the average of the two cells above it in the matrix.
3. When Bourgeois meets Bourgeois on half the occasions it is owner and wins while on half the occasions it is intruder and retreats. There is never any cost of display or injury.

In this game, Bourgeois is an ESS. If all the population are playing this strategy, no one ever engages in escalated fights because when two individuals contests for a resource, one is owner and the other is intruder; the result id that the intruder always gives way. With everyone playing the Bourgeois strategy the average pay-off for a contest is = 25. This is stable against invasion by Hawks, who would only get + 12.5, and also stable against Doves, who would only get =7.5. In fact Bourgeois is the only ESS in the game in Table 9.2 if all the population played Hawk, both Dove and Bourgeois could invade and do better. If all the population played Dove, then both Hawk and Bourgeois could invade.

Simple Models and Reality

These models are so simple in comparison with what actually happens in nature that we may well ask what use they can possibly be in helping us to understand real animal contests. There are three main conclusions.

1. The most important point is that the best fighting strategy for any one individual must depend on what other competitors are doing, because the pay-offs for employing a strategy will be frequency dependency dependents. Is Hawk a good strategy? The answer is yes if the population consists mainly of Doves and no if the population consists mainly of Hawks. Instead of asking whether a strategy is a good strategy, we should really instead be asking is it a stable strategy, or ESS?
2. The ESS will depend on the strategies in the game. In the first example, where there were just two strategies, the ESS was a mixture of Hawk and Dove. However, when we introduced another strategy, Bourgeois, into the game the ESS solution changed: Bourgeois turned out to be a pure ESS.

   These three strategies are undoubtedly too simple to represent in detail the strategies that animals adopt in the wild. Nevertheless, they are plausible alternatives which we could regard as simplified versions of strategies we might see in nature. It is interesting to find that neither Hawk nor Dove is an ESS in our simple games, but some mixture of Hawk and Dove behaviour can be stable. This is exactly what we see in real animals, a mixture of display and fighting. Because of its simplicity, we don't expect to actually see used simply for gaining some insights into how contest behaviour might evolve.

More detailed models will have to consider a greater variety of more complicated strategies and we will have to rely on biological institution to help us what strategies will evolve, only what will be stable given a defined set of alternatives. There is no doubt that an animal with a machine gun would invade the Hawk-Dove game and soon spread, but until we see real animals behaving like this there is little point in putting the strategy into our models.

3. The ESS will also depend on the values of the pay-offs in the game. If we changed on the pay-offs in Table 5.1 then the ESS mixture of Hawks and Doves would change. In fact, in this game it can be shown that as long as the cost of injury exceeds the value of winning then the ESS is a mixed one. The relative amount of Hawk-like behaviour we would expect will depend on the costs and benefits in the contest. If the value of the resource was greater than the cost of injury then pure Hawk is an ESS.

Therefore, if game theory models are going to make precise predictions about contest behaviour, we will not only need to know the range of possible strategies but also the costs and benefits for the pay-off matrix. In the real world, ecological circumstances such as resource abundance and competitor density will determine the pay-offs in the evolutionary game. In practice it is not going to be easy for the field worker to go out and measure on a common scale (effect on fitness) the cost of displaying, the cost of serious injury and the value of winning. However, we can see in an intuitive way that fighting strategies in nature vary depending on the value of the resource.

Examples of Animal Contests

Serious Fights

Convectional display does not always occur; sometimes fights are fierce and there is injury or death. Some weapons, such as horns and antlers, have evolved in relation to their efficiency in attack and defence. In the musk ox, from 5 to 10 percent of the adult bulls may die each year from fights over females (*Wilkinson* and *Shank*) and figures for mule deer indicate that up to 10 percent of male more than 1.5 years of age show signs of injury each year. Narwhals (*Monodon monoceros*) use their tusks for fighting and in one study over 60 per cent of the adult males had broken tusks;

some had tusk tips embedded in their jaws and most adult males were covered in head scars (*Silverman* and *Dunbar* 1980). Smaller animals may also fight viciously. Some male figwasps have large mandibles which are able to chop another male in half. When several males occur inside the same fig fruit they may engage in lethal combat for the opportunity to mate with females.

One such fruit was found to contain 15 females, 12 uninjured males and 42 damaged males who were dead or dying for fighting injuries. Damage included legs, antennae and heads completely bitten off, holes in the thorax, and eviscerated abdomens. In all these examples the value of the resource probably exceeds the cost of injury. We would therefore expect Hawk like strategies because failure in a contest could mean failure to pass on genes to future generations; individuals are really fighting for genetic life or death. Where, on the other hand, the resource is less valuable, or where the costs of fights are very high, then would we expect *Bourgeois* (owner wins–intruder retreats) to evolve as a strategy for settling contests?

Respect for Ownership

One possible examples comes from a study of lions by *Packer* and *Pusey* (1982). A coalition of several males (up to 7) defends the pride. Sometimes the males are close relatives, often brothers, bit 42% of the coalitions contain unrelated males. Competition among males within the pride for access to oestrus females consists of competition for temporary ownership of a female. When a female comes into oestrus, one male forms a consortship with her, guarding her from the approaches of other males and copulating with her frequently for a period of about 4 days. What determines which males get consortship with a particular female?

Contests between unrelated males were no more intense than between related males. Two important factors in settling contests were age and size; small males and very old or very young (less vigorous) males were less likely to gain access to an oestrus female.. Among prime males, however, the main factor influencing who won a contest was simply respect for ownership. Usually the first male to encounter an oestrus female guarded her and the other males then gave way. Most serious fights occurred when ownership was undecided, for example, when two males simultaneously came into the vicinity of an unconsorted and potentially oestrus female, or when ownership was unclear, for example, when the consorting male moved further form the female than a rival male. There was then usually a race between the males to arrive first at a female.

On arrival the loser would defer to the winner, even though in subsequent consortships the roles might be reversed with the previous consorting male now a rival to the new consort. Lions may use this Bourgeois-like strategy to settle contests because the cost of serious fight are so high and the benefits relatively low. Another example where ownership settled contests is a study of competition for territories among male speckled wood butterflies.

Observations showed that contests were brief and always won by the owner, and experiments showed that in a contest between two individuals the outcome could be reversed simply depending on who was the owner at the time. Finally, when two contestants were tricked experimentally into simultaneous ownership of a territory, an escalated contest ensued. In this last experiment the asymmetry of owner versus intruder was absent and both contestants behaved like Hawks. In another study, however, (*Wickman* and *Wiklund* 1983) contests were longer and appeared to involved assessment, with intruders sometimes beating owners. Like the lions, therefore, contests may sometimes be settled by fighting ability and sometimes by ownership.

(c) The influence of Resource Value

The observation that owners tend to win contests versus intruders does not, of course, necessarily mean that the Bourgeois strategy is being adopted. Owners my win for at least three reasons.

Hypothesis 1. They are better fighters. This may be the reason why they got ownership in the first place!

Hypothesis 2. They have more to gain from a fight and so are prepared to fight harder. If you were attacked in the street you would probably fight harder if you had a lot of money in your pocket than if you had very little. Similarly, owners may value the resource more than introders. For example in disputes over a territory the owner will be familiar with the territory the owner will know the location of the good feeding and nesting sites.

Hypothesis 3. Arbitrary asymmetry of ownership settles the contests. Ownership is simply a conventional settlement, as in the Hawk-Dove-Bourgeois game, similar to the conventions we use in deciding positions in a queue at a supermarket or occupancy of a seat on a bus.

Krebs (1982) tested these three hypotheses in a study of territorial defence by male great tits *Parus major.* He removed owners and kept them in cage. Newcomers took over the vacancy and then, after various time intervals the original owner was rereleased onto the territory. The prediction from hypothesis 1 is that the original owner will win back his territory, through his superior fighting ability. Hypothesis 2 predicts a gradual reversal of dominance in favour of the newcomer to learn the territory characteristics and so gain greater value from winning a contest. Hypothesis 3 predicts that the newcomer will win simply because it is now in the role of owner. The data supported the second hypothesis, with newcomers escalating and being more likely to win after they ad time to learn the territory characteristics and settle boundaries with the neighbours.

(d) Contests of Strength

So far we have considered, in a general way, how resource value and the behaviour of other individuals will influence fighting behaviour. However, as we saw from the lion example another important factor is the fighting ability or strength of the various contestants. When an individual fights for a resource its decision on how to fight must depend not only on the value of the prize but also on its own strength compared with that of its opponent. Animals in a population will differ in their strength and the costs of fighting a large individual will certainly be greater than those fighting a small one. Therefore, we might expect displays to sinal fighting ability and so allow the contestants to settle their dispute quickly without recourse to a costly fight.

Furthermore, displays used in assessment should be reliable signals of strength, otherwise weak individuals would be able to mimic the signal and gain an advantage. In the autumn, red deer stags (Cervus elaphus) compete with each other for females. Reproductive success depends on fighting ability; the strongest stags are able to command the largest harems and enjoy the most copulations. Although fighting brings great potential benefits, it also entails serious costs.

Almost all males suffer some slight injuries, and between 20 and 30 percent of stages will become permanently injured sometime during their lives, through broken legs or being blinded by an antler point for example. Competing stags minize fighting costs by assessment of each other's fighting potential and so avoid contests with individuals they are unlikely to beat. In the first stage of the display, the harem holder and challenger roar at each other. They start slowly at first and then escalate in rate. If the defender can roar at a faster rate then the intruder usually retreats. Roaring is a good signal of fighting ability because to roar well a stage has to be in good physical condition. At the peak of the rut, harem holders may be roaring at intruders throughout the day and night and, because they do not feed much during this time, they show a steady decline in body weight. Some stags literally become 'rutted out' and are unable to roar strongly. Their females may then be taken over by other stags.

In the second stage of a contest if a challenger outroars the defenders, or matches him, then he approaches and both stags engage in a parallel walk. This presumably enables them to assess each other more closely. Many fights end at this stage, but if the contestants are still equally matched then a serious fight ensues where they interlock antlers and push against and push against each other. Body weight and skillful footwork are important determinants of victory, but there is a chance that even the winner may get injured. The important point is that these escalated fights are rare, and most contests are settled at an earlier stage by displays. Many animal contests proceed like this and are direct or indirect trials of strength. Buffalos charge at each other and assess their fighting potential in head-on clashes. Beetles engage in pushing contests in which the larger one emerges as victor. Male frogs and toads have wrestling matches where the large ones win the best territories or the most females. In many species of frogs and toads the pitch of a male's croak is related to his body size; the larger the male, the larger the vocal cords and so the deeper the croak.

Common toads (*Bufo bufo*) assess the body size, and hence fighting potential, of their rivals by the pitch of their croaks. An attacker is much less likely to attempt a take-over of a female when deep croaks are broadcast from a loudspeaker next to the pair than when high pitched croaks are played. Often both strength and ownership are cues used to settle contests. When crabs compete for burrows or when spiders compete for webs the larger individuals usually win the fights, but if the contestants are equal in size then it is often the owner who wins. In conclusion, the game theory models we discussed earlier are too strategies but, rather, will adopt strategies conditional on the fighting potential of their opponents.

Contests with Differences in Both Resource Value and Fighting Ability

So far we have identified both resource value and fighting ability as important factors influencing the outcome of contests. In many contests the opponents will vary both in fighting ability and how much they value the resource. How will fighting strategies evolve in these cases?

Theory: The Asymmetric War of Attrition

The question of whether we expect to see contests in nature settled simply by the arbitrary convention of ownership, as opposed to real difference in fighting ability, depends on how individuals can vary their degree of risks of injury in a fight. In the Hawk-Dove-Bourgeois game, contestants can engage in conflicts at only two levels of risk, namely display (low risk) and escalated fighting (high risk). This discontinuity is crucial for the generation of the Bourgeois strategy as an 'uncorrelated' asymmetry to settle contests. (Uncorrelated means ownership need have no correlation with fighting ability or resource value, for example; it is simply an arbitrary convention to settle disputes without escalation).

If the contestants can control their degree of risk along a continuum, however and therefore adjust their escalation in relation to the value of the resource, then the uncorrelated Bourgeois rule turns out to be no longer evolutionarily stable. A contest in which there are differences in fighting ability, and in which individuals can vary their cost of escalation along a continuum is referred to as an 'asymmetric war of attrition'. The ESS for this kind of contest is for individual A to give up and retreat when

$$\frac{V_A}{K_A} < \frac{V_B}{K_B}$$

where V is the value of the resource to contestants A and B, and K is the rate at which the two individuals accrue costs during a contest. K will be related to fighting ability; good fighters will accrue costs at a slower rate than poor fighters.

Therefore if ownership is used to settle contests of the war of attrition kind, it is because there are asymmetries in V and / or K which favour residents, and residents are willing to escalate further. A verbal summary of this rule for settling contests is withdraw if you will run out of fitness budget before your opponent.' This is the common sense solution to the game; the winner is always the one with the greatest benefit: cost ratio.

Data: Contents Between Male Spiders

Steven Austad (1982, 1983) studied for females between male bowl and doily spiders, *Frontinella pyramitela.* One mature, the males feed very little and spend most of their time going around female webs in search of mating opportunities. They only live for about 3 days in the final stage and so the way they compete for females is vital for their reproductive success. Laboratory experiments showed that there was first male priority in sperm competition. These experiments involved allowing two males to mare with a female, one normal male and one irradiated male. The eggs fertilized by the normal male developed normally while the eggs fertilized by the irradiated male failed the develop because of genetic abnormalities induced by the irradiation. If the sequence of matings was first the normal, then the irradiated male, only 95% of the eggs developed (therefore 95% were fertilized by the first male). If the sequence was first the irradiated and then the normal male, only 5% of the eggs developed (therefore 95% were again fertilized by the first male).

Note that this is the reverse of what happens in dungflies, where the *second* male has the advantage. The second finding was that there was no sperm displacement as a result of a take-over by another male; if the first male had transferred enough sperm to fertilize 30% of the female's eggs he would fertilize these even if another male then took over and copulated. This means that if a male lost a fight half way through the mating, he would lose only future possible gains; the take-over would not affect his past success.

(a) Measuring Resource Value, V

Copulation proceeds as follows. First there is a pre-insemination phase during which the male assesses the female, for example to discover whether she is sexually mature. Then copulation itself occurs, the male passing sperm from his pedipalps to the female. The proportion of eggs fertilized increases at a rapid rate early on in copulation and then at a slower rate as time goes on. The result is that the female is of different value to the male at different stages of the mating sequence. The average number of eggs fertilized, for all females meets encountered by a male, is 10 eggs, so when a male first meets a female her expected value is 10.

Imagine this female is in fact a virgin, which is the most profitable female that a male encounters, being worth on average 40 eggs. During the pre-insemination phase the value increases to 40 as the male discovers that the female is a virgin.

Copulation then begins and the value of the female declines as her eggs are fertilized. After 7 min of copulation, 90% of the eggs have been fertilized so the female is now worth only 4 eggs. After 21 min of copulation, the values is less than 1 egg because 99% of the eggs have now been fertilized. Imagine now that an intruding male comes along at different stages in this mating sequence by the resident male. Intruders are unaware of the state of the female unless they themselves are able to assess her.

Therefore the intruder will always assume that the females is of average value, 10. This meant that Austad could vary V in a fight simply by introducing intruders at different stages. At the end of the pre-insemination phase, for example, V is higher for the resident who has the knowledge that the female is a virgin and particularly valuable. After 7 min of copulation, V is higher for the intruder, because only the resident has the knowledge that most of the eggs are already fertilized. We may expect residents to fight harder for the female are the end of the pre-insemination phase than after 7 min of copulation.

(b) Measuring the Costs of a Fight, K

This is the expected reduction in life time reproductive success per minute of fighting. Austad found that costs, measured in terms of the probability of serious injury (which is a measure of loss of future reproductive success), were linearly related to the duration of the contest. In fights, the males grab each other and grapple with their jaws and legs. Long fights led to body wounds, loss of legs and eventually certain death. A male's fighting ability depended on his body size. Therefore, it was possible to vary K by staging encounters between males of different size. Many contest were arranged, varying V and K as described above. The results were as follows:

1. When V was the same for both males (both introduced simultaneously at the start of the pre-insemination phase) fights were settled by difference in K. Larger males won 82% of the fights and grapple duration increased with closer matching of body size between the contestants presumably because it then took longer to assess difference in fighting ability.

2. When K was the same (same size opponents), fights were settled by differences in V. Thus residents fought for longer and were more likely to win at the end of the pre-insemination phase, whereas they fought for less time and intruders were likely to win at the stage of 7 min of copulation.
3. When residents were smaller than intruders, they persisted more in fights when V was greater. For example, at the end of the pre-insemination phase, when V was at its highest for the resident, small residents would persist in fights for so long that 90% of the grapples led to them sustaining serious injury. After 7 min of copulation, however, they gave up much sooner and only 30% of the fights led to injury.
4. The most serious escalations occurred where V/K was indentical for both contestants. In these cases neither male was prepared to give way and most of the fights ended in serious injury and death for one of them. This occurred, for example, in fights between small residents versus large intruders at the end of the pre-insemination phase. Here V and K were both large for the resident, and were both small for the intruder.

Austad's experiments show clearly that both V and K influence duration of contests and who wins. Similar results have been obtained by *Sigurjonsdottir* and *Parker* (1981) who studied struggles between male dungflies for the possession of females on cowpats.

Fighting for Dominance in Flocks

Some animals live in groups and individual differences in fighting ability determine who will have priority access to food or males. Many displays within groups probably involve reliable assessment of strength; however, the exact link between the display and fighting ability is not always obvious. In the Harris sparrow (*Zonotrichia querula*) there is enormous plumage variability in winter when the birds go around in flocks searching for food. Individuals with the blackest plumage are the dominants and they always displace pale birds from food supplies. If blackness signals dominance than why don't the subordinates grow blacker feathers and enjoy a rise in status?

In the spring all the birds come into breeding plumage and even the pale males develop black colouration, so the dark feathering is not an obviously uncheatable signal of strength like the roar of a rutting stag or the deep croak of a large toad. The Rohwers attempted to create 'cheats' in the flock by experimental treatment of subordinate birds. In the first experiment they simply painted subordinates black. These treated birds were attacked by the others and failed to rise in status. Secondly some subordinates were injected with testosterone but their pale plumage was left unaltered. These birds behaved more aggressively and attempted to assert their dominance but they did not rise in status because their opponents did not retreat during disputes. Finally, subordinates were painted and injected so that they both looked and behaved like dominants. This time the cheating worked: the birds won more fights and were respected by other in the flock.

Table 9.3. Summary of the Experiment on Signals of Dominance in Flocks of the Harris Sparrow

Experimental treatment of subordinates	*Look dominant*	*Behave as dominant*	*Rise in status?*
1. Paint black	Yes	No	No
2. Inject testosterone	No	Yes	No
3. Paint black and inject with testosterone	Yes	Yes	Yes

Therefore, the earlier attempt to create a cheat by planning alone did not fail simply because the other birds didn't like the paint! It failed because although the darkened birds looked dominant, they did not behave so. The conclusion is that plumage alone is not a passport to high status in the flock; the signal must be packed up by dominant behaviour as well. Nevertheless, the dark plumage still acts as a badge to reduce the amount of fighting in the flock. In another experiment dominant birds were beached so that they signalled low rank. They were attacked a lot by others who exerted their superiority but only after a lot of squabbling. It is clear from the experiments above that cheating is prevented because a subordinate cannot become dominant through darker plumage alone. But why can't a low ranking bird increase its testosterone level as well? This same question arises from work on red grouse where males were injected arises testosterone and immediately doubled their territory size increased the number of females they attracted.

Presumably the answer to this problem is that an increase in androgen levels would push the behaviour of a subordinate bird beyond its real capabilities. Although it may gain a short term increase in success, perhaps the long term effect would be that all its strength is sapped and it suffers a decrease in fitness. Not all cases of plumage variability are concerned with status signalling. For example, a wading bird, the turnstone (*Arenaria interpres*) has very variable plumage. *Whitfield* (1986) removed to either resemble the removed owner or to be lighter or darker. Neighbours were much less likely to attack the models if they looked like the removed owner than if they looked different. However the attack intensity towards models that looked different was nor related to whether the models were light or dark. In this species, plumage variability may simply facilitate individual recognition.

INFORMATION TRANSMISSION AND ANIMAL CONTESTS

During a contest there are two sorts of information that a display could signal to an opponent.

Information about Strength

For example '1 am 2 m tall'. This kind of information is often present in animal displays. It is difficult to fake and we have already encountered examples, such as roaring and deep croaks, which are reliable signals of fighting potential.

Information About Intentions

For example, 'I am gong to attack you' or 'I am gong to display for 3 min.' This kind of information will be easier to fake and game theory predicts that it will not be transmitted in displays. Returning to the war of attrition model described at the beginning of this chapter, imagine that individual A choose a cost m_A which is larger than the cost m_B chosen by B. Would it not pay to signal their intentions at the start of the contest? B would then retreat at once and both contestants would be better off; A would get V instead of $V-m_B$ and B would get O instead of $-m_B$. The problem with signalling intentions is, of course, that of cheating. In a population of honest signalers, it would pay an individual to lie and signal a higher cost so that it could win every time. Eventually every one would be lying and it would no longer pay to believe the message. Mutants which ignored the signals and just obeyed the war of attrition distribution would do better.

The end result, the ESS, is for no one to give any information about intentions and for all to display according to the war of attrition model. As Maynard Smith point out, all this really means is that you should not believe what an opponent at poker tells you! The prediction from this line of argument, that displays should not evolve to transmit information about intentions, is to a large extent borne out by the ethological literature. In blue tits (*Parus caeruleus*) a particular display can be followed by several different types of behaviour and there is no one posture that is followed with a high probability by attack.

In the Siamese fighting fish (*Betta splendens*) there is no consistent difference between the behaviour of the eventual winner and eventual loser until near the end of the contest, so it is impossible to predict which individual will win from its displays. The evidence suggests, therefore, that information about intentions is not transmitted. However, there are two lines of research that need further investigation. Firstly, it is clear that animal use a great variety of displays in agonistic encounters and at present game theory has little to say about how the variety might have evolved.

Secondly, although it is clear that no one display is a good predictor of future behaviour, sequences of displays may be important. Film analysis of *Muhammed Ali* in the boxing ring shows that about half of his left jabs are of shorter duration that the visual reaction time. Nevertheless, his opponents still managed to evade most of the punches. They could not have done this by responding to the preliminary stages of the jab. Somehow, therefore, they must have decoded Ali's behavioural sequences so as to predict the punch before it actually starts its movement.

Fighting

When an animal is cornered by a predator, it will often fight. This type of fighting, the defence against a predator, will not however concern us here, because it usually does not involve animals of the same species. Nor is it as common as the fighting of animals which is directed at individuals of their own species. Most of this intraspecific fighting is done in the breeding season, and is therefore called reproductive fighting. Some fighting has to do with dominance relationships in the group and is not linked with the breeding season.

Reproductive Fighting

Different species fight in different ways. Firstly, the weapons used are different. Dogs bite each other, and so do gulls, and various fish. To that end alone the male Salmon develops a formidable jaw. Horses and many other hoofed animals try to kick each other with the forelegs. Deer measure their relative strength by pushing against each other with their antlers. Waterhens can be seen to fight all through the spring in many parks. They throw themselves halfway on their backs, and fight with their long-toed feet. Many fish fight by sending a strong water-jet towards the opponent by means of vigorous sideways tail-beats.

Although they do not actually touch each other, the movement in the water caused by the tail-beats gives a powerful stimulus to the opponent's highly sensitive lateral-line organs. Male Bitterlings develop horny warts on the head in spring, and try to butt each other with the head. Secondly, although so much fighting goes on all through the spring, it is relatively rare to see two animals actually engaged in "mortal combat" and wondering each other. Most fights take the form of "bluff" of threat. The effect of threat is much the same as that of actual fighting: it tends to space individuals out because they mutually repel each other. Its variety is almost endless. Great Tits threaten by facing each other, stretching the head upward, and swinging slowly from side to side, thus displaying the black-and-white head pattern. Robins threaten by displaying the red breast, turning it to the opponent and then turning slowly right and left alternatively.

Some Cichlids display the gill covers by raising them while facing the enemy. In *Cychlasoma meeki* and in *Hemichromis bimaculatus* these gill covers are adorned with very marked black spots, bordered by a golden ring; the threat display shows them off beautifully. Not all threat is visual. Many mammals deposit "scent signals" at places where they meet or expect rivals. Dogs urinate to that purpose; Hyaenas, Martens Chamois, various Antelopes and many other species have special glands, the secreta of which are deposited on the ground, on bushes, tree stumps, rocks, etc. The Brown bear rubs its back against a tree, urinating while it does so. Sounds may also have a threat function. All the calls already mentioned under the collective heading "song", do not merely attract females, but serve to repel males as well.

The Functions of Reproductive Fighting

Reproductive fighting is always aimed at a special category of individuals. In most species it is the males which fight, and they attack exclusively, or mainly, other males of the same species.

Some times male and female both fight; when that happens there is often a double fight; male attacking male, and female fighting female. In the Phalaropes and some other species of birds it is the females who fight, and they again attack mainly other females. This all shows that fighting is aimed at reproductive rivals. Further, fighting, and threat as well, tend to prevent two rivals or competitors from settling at the same spot; mutual hostility makes them space out, and thus reserve part of the available space for themselves. An examination of what is essential in this reserved space will help us understand the significance of fighting.

The fighting of each individual is usually restricted to a limited area. This may be the area round the female(s) as it is in Deer and many other animals. Bitterling males defend the area around a Freshwater Mussel against other males to this Mussel they attract a female. They induce her to lay her eggs in its mantle cavity, where they will develop, leading a parasitic life. Carrion Beetles of the genus *Necrophorus* defend carrion against rivals. The defence in all these cases not only concerns the central object itself, but also a certain area round it; rivals are kept at a considerable distance. In the species mentioned it is easy to see what the central object is : when a doe moves and walks off the male will go with her; it always fights in her vicinity. When the Mussel moves, the male Bitterling shifts the area it defends along with it. In most species however the defended area does not move; the male settles down on a chosen spot, and defends a territory. This is known in many animals; territorial fighting and threat can be watched in every garden, for Robins, Chaffinches and Wrens, to mention only a few species renowned fighters.

It is possible to understand the significance of such a territory when the fighting is centered on one particular part of it. Thus in many hole-breeding birds the fighting is particularly furious when intruders come near the hole. In most species, however, there is no such concentration on a particular part of the territory, and here the significance of territory is less easy to understand. It has been suggested that the territory of many songbirds might be useful as a reservoir of food for the young. This would enable the parents to collect a certain basic quantity of food near the nest, which would mean that the foraging trips could be short. Since newly hatched songbirds have to be brooded in order to remain warm and ready to gape for food, the territory might be a help in keeping foraging trips, and thus intervals between bouts of brooding, as short as possible.

The length of the intervals between bouts of brooding might, on unfavourable days, be critical. Opinions about the value of this argument differ, however. In ground-breeding such as Gulls, Terns, Lapwings, etc., spacing-out appears to be part of the defence of the brood against predators. There is evidence to show that too dense a concentration of prey such as eggs of chicks makes predators specialise on them; that is the main reason why camouflaged animals are as a rule solitary and well spaced out. In birds like Gulls, where the brood is camouflaged, territorial fighting has the effect of keeping the individual broods reasonably far apart.

Here again the conflict between two interests has resulted in compromise: social nesting has certain advantages; so has spacing-out. The various species of Gulls and Terns have each arrived at a compromise which give them some, though not complete, benefit of both tendencies. Concluding, it is clear that reproductive fighting serves a function. It results in a spacing-out of individuals, thus ensuring each of them the possession of some object, or a territory, which is indispensable for reproduction. It thus prevents individuals sharing such objects, which would in many cases be disastrous, or at least inefficient. Too many Bitterling eggs in a Mussel will result in a law ration for each. When many males would mate with one female instead of securing females of their own this would be a waste of germ cells. Two broods of Starling in one hole may be fatal to both broods. Spacing out makes the individuals utilise the available opportunities.

The Causes of Fighting

One next problem is: what makes the animals fight in such a way as to promote these functions? What makes them fight only when it is necessary, and only at the place where it is necessary? How does the animal select its potential rival amongst the multitude of other animals it meets? Since fighting endangers the individual (because it makes it vulnerable to attack by predators), and since it may endanger success in reproduction because unlimited fighting might leave the animal little time to do anything else, restriction of fighting to situations in which it can serve its functions is of vital importance. These problems are rather similar to those discussed in relation to mating.

In order to confine fighting to the actual defence of territory, the Mussel, the female, etc., the animal will have to react specifically to these situations. Further, fighting must be timed, that is, confined to those moments when there is a rival to be driven off. Finally, it must not be wasted on other species, except when these are rivals. As we will see, many of the outside stimuli responsible for these various aspects of co-ordination are provided by the rival. Since moreover, most of these stimuli serve more than one of these functions. I will not divide this treatment into sections according to the function served as rigidly as I did in the chapter on mating. As we have seen, restriction to a certain locality is one of the most obvious characteristics of fighting. When a male Stickleback meets another male in spring, it will by no means always fight. Whether it does depend entirely on where it is. When in its own territory, it attacks all trespassing rivals.

When outside it territory, it will flee from the very same male which it would attack when 'at home.' This can be nicely demonstrated in an aquarium, provided it is large enough to hold two territories. Male A attacks male B when the latter comes into A's territory; B attacks A when A trespassers. Usually neither male trespasses voluntarily on to strange territory, but one can easily provoke the situation by capturing the males and putting each of them in a wide glass tube. When both tubes are lowered into territory A, A will try to attack B through the double glass wall, and B will frantically try to escape. When both tubes are moved into territory B, the situation is completely reversed. How the territory stimulates the male to fight has rarely been studied in any detail. It can of course only be found out by experimentally moving the territory or parts of it and seeing whether the male adjusts its fighting to the changed situation. This is of course difficult in birds because of their large territories, but small fish that can be kept in aquaria offer unique opportunities for study.

Several cases have been reported of birds extending their territories after the female had started to build a nest outside the territory originally staked out by the male. It seems certain that territories are selected mainly on the basis of properties to which the animals react innately. This makes all animals of the same species, or at least of the same population, select the same general type of habitat. However, the personal binding of a male to its own territory–a particular representative of the species breeding habitat–is the result of a learning process.

A male Stickleback is born with a general tendency to select a habitat in shallow water with liberal vegetation, but it is not born with the tendency to react to a particular plant here and a pebble there. It shifts its territory when we move these landmarks because it has been conditioned to them. This can be seen from the fact that when it breeds two or three times in succession, it often moves to new territories. In each of them it orients itself to landmarks. Species which react to a special object, such as a hole, or as the Bitterling does, to a Mussel, probably react innately to it and as a consequence react to only few "sign stimuli" provided by it. The Bitterling, for instance, reacts only to a minor extent to the visual stimuli provided by the Mussel; the fish reacts both to the movement of the water and to its chemical properties. Stimuli from the territories to which the animal reacts either innately or as an added result of conditioning, makes the animal confine its fighting to the territory.

The gross timing of the attack is again a matter of outside factors. As in mating, the first, very crude timing depends on sex hormones. Fighting appears as a consequence of gonadal growth which, in its turn, through the pituitary gland, depends or rhythmic factors such as day lengthening in the case of many animals of the northern temperate zone. The more accurate timing however is again a matter of reaction to signals. Signals from the rival releases fighting when the latter comes too near the territory, or whatever the defended object may be. These signals always have a curious double function. When displayed by a stranger, they draw the attacker to it. When displayed by an attacker or its own ground, they intimidate the stranger. When experimenting with models one can release both responses, depending on the place–inside or outside the territory–where they are presented.

In both cases they serve to space out the species, and, since the responses are specifically released by these displays, not by threat displays of other species, they tend to confine hostilities within one species. The stimuli have been analysed in various species by experiments with models. The male three-spined Stickleback, while showing some hostility towards any trespassing fish, concentrates on males of its own species. Models of males release the same response, provided they are red underneath. A bright blue eye and a light bluish back add a little to the model's effectiveness, but shape and size do not matter within very wide limits.

A cigar-shaped model with just an eye and a red underside releases much more intensive attack than a perfectly shaped model or even a freshly killed Stickleback which is not red. Size has so little influence that all males which I observed even "attacked" the red mail vans passing about a hundred yards away; that is to say they raised their dorsal spines and made frantic attempts to reach them, which of course was prevented by the glass wall of the aquarium. When a van passed the laboratory, where a row of twenty aquaria were situated along the large windows, all males dashed towards the window side of their tanks and followed the van from one corner of their tank to the other. Because models of three times stickleback size, although releasing a similar attack as long as they were not too close, were not actually attacked when brought into the territory, it seems that the angle subtended by the object is important and this must be the reason why the distant mail vans were attacked. Apart from colour, behaviour may release the attack.

A male-Stickleback viewing a neighbour from a far often adopts a threat posture, a curious vertical attitude, head downward. The side, or even the underside, is turned toward the opponent, and one or both ventral spines are erected. This posture has an infuriating effect on other males, and we can intensify a male's attack on a dummy by presenting it in this posture. Similar observations have been made on the Robin. When a male Robin has staked out his territory, the sight of another Robin in this territory releases attack or threat. Lack showed that the red breast is the releasing factor more than anything else. When he put up a mounted Robin in an occupied territory, this was postured at by the owner.

Even a small cluster of red feathers was sufficient to evoke posturing. And, just as in the Stickleback a very crude red model was more effective than a perfectly shaped but silvery model, so to the Robin these few red feathers had more meaning than an entire mounted immature Robin, which had all the form-feathers of its species but had a brown instead of a red breast. It is remarkable how similar are the functions of the red breast of the male Stickleback and those of the Robin's red breast. We will see that comparable signalling systems have convergently developed in animals of other groups as well. In the Robin, the signalling is not entirely visual however. Robins hear each other over far greater distances than they see each other.

In particular, the song of a Robin arouses the owner of a territory and sets him off to find the singer. The real attack therefore occurs in at least two steps; a male first flies in the direction from which it hears another male singing; it then looks round and is roused by the red breast of the intruder to posture or attach. In many other species song has this same function; it is a "badge of masculinity" and releases fighting in territory holders.

As stated already that such a badge of masculinity, when demonstrated by a male-on its own territory, drives off intruders. This can easily be seen in the field without experimentation, because it often happens that a singing male cannot be seen, when for instance it is hidden by a tree or a shrub. It is fascinating to watch the intense reactions of other birds to such concealed singers. Trespassers are personifications of a bad conscience; territory owners those of righteous indignation. In the Herring Gull, though the sexes have the same coloration, aggressiveness is mainly found in the males, and directed against other Herring Gull males.

Male Herring Gulls however do not sing, nor does any of their calls particularly arouse other males. Nor do the males have conspicuously coloured parts that release fighting in other males. Their behaviour however does. Threat postures and nest building movements particularly draw the attention of other males and elicit a hostile response from them. Other species again resemble the Stickleback in that the male's bright colours act as their badge. This for instance was found to be the case in the American Flicker (*Colaptes* auratus), a Woodpecker in which the male has a patch at the corner of the mouth (the so-called moustache) which the female has not. When the female of a pair was captured and given an artificial black moustache she was attacked by her own mate. After recapture and removal of the moustache she was again accepted. Male Shell Parrakeets (*Melopsittacus undulatus*) differ from females in the colour of the cere, which is blue in males, brown in females. Females with the cere painted blue were attacked by males. A most remarkable parallel was found in a group as different as the Cephalopods.

Males of the Common Cuttlefish, *Sepia officinalis*, have a brilliant visual display at the mating time. On meeting another Sepia they show the broadside of their arms, and by co-ordinated action of their chromatophores develop a conspicuous pattern of very dark purple and white. The fighting of the males is a response to this male display; experiments with plaster models showed that the display acts visually; both shape and colour pattern contributing to the release of attack. Lizards behave much as Cuttlefish do. The males have special movements which serve to display specific male colours. The American Fence Lizard (*Sceloporus undulatus*) has procryptic colours on the back. The underside of the males, however, is a clear blue. This colour is not visible until the male displays, as it does in spring upon meeting another Fence Lizard. It then takes up a position in front of the other and at right angles to it, and compresses its body laterally so that the blue underside becomes visible from the side.

By changing the colour of males and females with lacquers, Noble showed that the blue belly releases fighting in territory holding males. In most of these cases they direct the fighting at the same time. However, as in mating behaviour, we have to distinguish between these two functions, for there are stimuli contributing to one and not to the other. In ducks, for instance, females make species movements and calls urging their mates to attack other males. The cells merely raise the male's aggressiveness, but by special head movements the female points out to her mate the male to be attracted. This can easily be seen in the tame and half-tame Mallards living in our parks: the female swims from an "accosting" male to her own mate, repeatedly pointing her head with a sideways movement over her shoulder in the direction of the stranger. The third problem, related to reproductive isolation, that of confining the fighting to members of the same species, has also been covered by the examples given. Again, as with the signals which play a part in mating behaviour, the fight-evoking signals are specific, and very different even in closely related species if they are living in something like the same habitat.

Yet one gets the impression that interspecific fighting is not as rigorously eliminated in evolution as interspecific mating. It seems, so far as the evidence goes, that what little interspecific fighting occurs is directed at species superficially resembling the species of the attacking animal. "Erroneous" attack occur because a strange species happens to present some of the sign stimuli normally releasing attack. In some cases, however, fighting is clearly aimed at other species because they are competitors for the same "indispensable object." Thus Starlings and Tree Sparrows are known to drive other species from nesting holes.

The Peck-Order

Animal species living in groups sometimes fight over other issues than females or territories. Individuals may clash over food, over a favourite perch, or possibly for other reasons. In such cases, learning often reduces the amount of fighting. Each individuals learns, Individuals learns, by pleasant or bitter experience, which of its companions are stronger and must be avoided, and which are weaker and can be intimidated. In this way the "peck-order" originates, in which each individual in the group knows its own place. One individual is the tyrant; it dominates all the others. One is subordinate to nobody but the tyrant. Number three dominates all except the two above it, and so on. This has been found in various birds, mammals, and fish. It can easily be seen in a hen-pen. The peck-order is another means of reducing the amount of actual fighting.

Individuals that do not learn quickly to avoid their "superiors" are at a disadvantage both because they receive more beatings and because they are an easier prey to predators during fights. The behaviour leading to peck-orders has some interesting aspects. Lorenz found in Jackdaws that when a female of low "rank" got engaged to a male high up in the scale, this female immediately rose to the same rank as the male, that is to say that all the individuals inferior to the male avoided her though several of them had been of higher rank than she before. The American literature contains many valuable contributions on the problems of peck-order. In many of these papers, however, peck-order is claimed to be the only principle of social organisation. The leads to distorted views; peck-order relationship from only one category among the numerous types of social relationships in existence.

10

POPULATION BEHAVIOUR

The natural regulation of animal populations has long fascinated ecologists, and the role of behaviour in this process has been the subject of much debate. A complex of abiotic substances (minerals and organic compounds), climate (mainly temperature, moisture and wind), and biotic factors (other organisms of the same and other species) affects the growth and reproduction of any organism. A *population* may be defined as a group of organisms of the same species in a particular place. We know that all species have a reproductive potential that is seldom realized and that no population increases without limit, as was pointed out by *Darwin* and by *Malthus* before him. The number of individuals present in a population is determined by four population *forces:* births, deaths, and movements out of (immigration) and into (immigration) an area. Thus, the abundance of a species as well as its distribution is affected by behaviour. This chapter briefly reviews the extrinsic limits to population growth and then considers the evidence for these suggested intrinsic processes of population self-regulation: (1) purely behavioural mechanisms. (2) behavioural-physiological mechanisms, and (3) behavioural-genetic mechanisms. We then discuss how such system could have involved by natural selection.

LIMITING FACTORS

One axiom of ecology is that only a single factor limits the growth of population at any one time. The idea originated with *Liebig* (1847), a plant physiologist who studied effects of nutrients, such as nitrogen, on plant growth and is referred to as Leibig's *Law of the minimum.* Extending his idea to animal populations, we can test which factor limits a population's growth if we manipulate the amount of one factor while holding other constant and record changes in the population. If food is the limiting factor, an increase in the population will result if the food supply is augmented. For example, *Holling* studied the population response of some small mammals to different amounts of a staple food cocoons of the European pine sawfly (*Neodiprion sertifer*). Shrews of the genus *Sorex* increased More or less linearly with increasing amounts of the food supply up to a population of twenty-five shrews per acre; then the population leveled off. At that point some resource other than sawflies became limiting.

Likewise, deermice (*Peromyscus*) increased in numbers up to about eight per acre, then remained constant despite the food supply. The abundance of saw flies seemed to be unrelated to the population density of another shrew of the genus *Blarina.* While Liebig's law of the minimum can simplify our analysis of regulatory factors, it does not help us understand the level at which these factors may operate. Ultimately, food supply usually sets the limit to the number of organisms that can live in place–that is, the *carrying capacity* of the environment.

However, some behavioural mechanism, involved with contest competition for food, such as defense of territory, may proximally limit the population below the carrying capacity if some individuals defend space in which to live and reproduce and thereby exclude others. Thus both territory and food supply may limit the

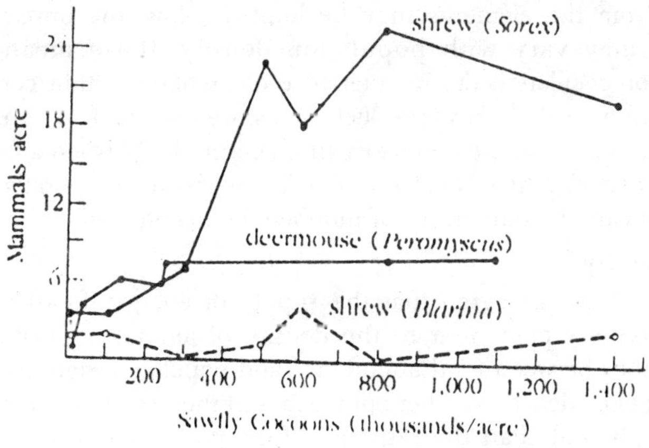

Fig. 10.1. Number of mammals per acre as a function of food supply.

population but operate at different levels, one proximate and the other ultimate. The interaction of factors creates more complicated problems; for example, plants that are deficient in nitrogen are more susceptible to drought, and most terrestrial insects can tolerate higher temperatures when humidity is high than when humidity is low. In addition, the impact of a limiting factor may occur during a very short period in the history of the population and thus be difficult to identify. A few days of severe winter weather may reduce a deer herd to the point that other regulatory factors never come into play. Therefore, it should not be surprising to find little agreement among population biologists about the mechanisms of population regulation.

Density-independent and Density-dependent Factors

Regulatory factors are sometimes divided into two types: density independent and density. *Density-independent* cause of mortality–such as climate, flood and fires–are not themselves affected by population density; they kill organisms regardless of density. *Density-dependent* factors–which include competition, parasitism, disease, predation and food supply–are related directly or inversely to density, so that as population density increases, death or immigration rates increase or birth rates decrease.

Actually, this division is some what misleading, since some of the effects of so-called density-independent factors may be modified by population density. At high densities, organisms may gain protection from both wind and low temperature by grouping, or shelter from the elements may be limited. Thus, the impact of climate may vary with population density. If we mean by population regulation the maintenance of numbers within certain limits, and not simply random fluctuations, we must look to density-dependent factors as the regulatory mechanisms. Only when a factor increases its negative effect on population growth as population increases can an equilibrium of numbers be maintained.

Food Supply

Most ecologist agree that the supply of energy (food) most often sets the upper limit to the density of animal populations. Correlations between food abundance and population density, are one type of evidence. Another approach is to supplement the natural food supply and chart the response of the population in terms of density, reproduction, mortality and movements. In one such study laboratory mouse-food pellets were distributed in a wood-lot and the response of the resident white-footed mice (*Peromyscus leucopus*) was noted. By the end of the study the population was somewhat higher than was that in the control area, primarily due to higher survival of young in fall and winter.

When oats were supplemented to populations of deermice (*Peromyscus maniculatus*) breeding began within several weeks, even in the winter, Home ranges decreased in size and immigration tripled compared to untreated control plots, suggesting that spacing behaviour may have been involved in limiting numbers. In spite

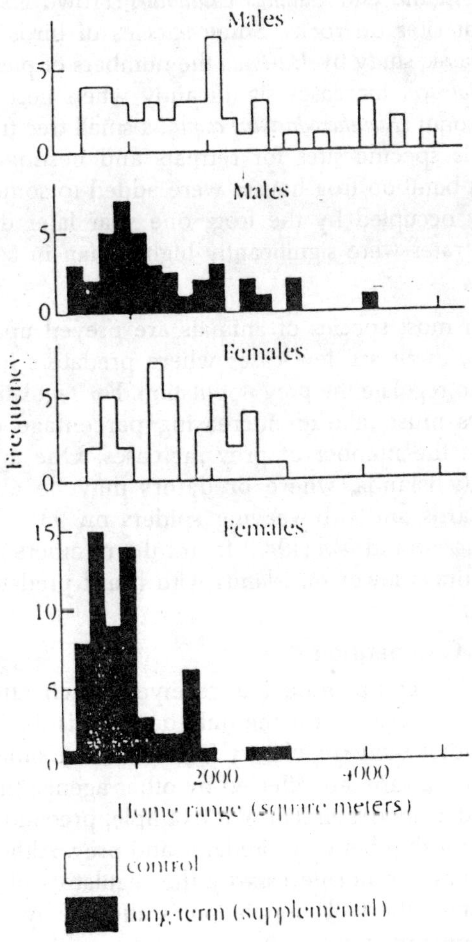

Fig. 10.2. Effect of supplemental food on home range size in deermice.

of the presumed importance of food as a limiting resource for vertebrate populations, naturalist have frequently noted that starving animals are rarely found in the wild, except where humans have altered the environment. These observations have contributed to the notion that behavioural mechanisms keep numbers below the point at which the food supply is exhausted.

Shelter

A few studies have demonstrated that the number of nesting or feeding sites limit population density. Filter feeders, such as sessile invertebrates, may compete actively for space. Barnacles (*Chthamalus stellatus* and *Balanus balanoides*) crowd each other out at attachment sites on rocks. Some species of birds nest in tree holes. In a classic study by *Hartman* the numbers of pied flycatchers (*Ficedula hypoleuca*) increases significantly when nest boxes were added. The coqui (*Eleutherodactylus coqui*), a small tree frog of Puerto Rico, defends specific sites for retreats and nesting in the rain forest. When bamboo frog houses were added to some plots, they were quickly occupied by the frog; one year later densities and reproductive rates were significantly higher than in control plots.

Predators

Although most species of animals are preyed upon by some other species, there are few cases where predators have actually been shown to regulate the prey population. For regulation to occur, the predators must take an increasing percentage of the prey population as the number of prey increases. One approach has been to study islands, where predators may be absent. After censusing lizards and orb-weaving spiders on 93 islands in the Bahamas, *Schoener* and *Toft* (1983) found the numbers of spiders to be about 10 times lower on islands with lizard predators than on those without.

Intraspecific Competition

Intraspecific competition has received much attention as a regulatory factor because it is the only one that is directly density-dependent. Other biotic regulatory factors involve other species of organisms that, in turn, are affected by other agents; thus, they are not perfectly density-dependent. For example, predation involves a complex relationship between predator and prey. Although wolves eat caribou, they are not necessarily the regulators of the caribou population. Since the wolf population is affected by factors other than the numbers of caribou, such as diseases or the abundance of other prey species, it will not be able to parallel the caribou population exactly. An outbreak of disease might reduce the size of the wolf pack and allow the caribou to *escape* control and begin to destroy grazing land. Competition among the caribou, however, will vary directly with the population density and the limiting resource.

Population Self-Regulation

The self-regulation school of thought argues that behavioural and physiological changes take place in animals because of competition with conspecifics. As a resource becomes scarce, the intensity of contest competition increases and causes changes within the organisms that increase the deaths and immigrations and/or decrease the birth rate. The adaptive value of this system is thought by some to be that the population is regulated below the limit set by the environment. If the limiting resource is food, over-exploitation could permanently lower the carrying capacity; for example, overgrazing causes soil loss from erosion by wind or water. Contest competition in advance of such a shortage could adjust birth, and movement rates and thus avoid wasted energy in reproduction. Although proponents of self-regulation have proposed many eloquent explanations, a big stumbling block is that most mechanisms seem to require individuals to sacrifice direct fitness by not breeding or by dying for the good of the rest of the population.

The evolution of self-regulatory mechanisms of population regulation appears to involve selection at the level of the group. To put the problem in human terms, let us consider the so-called tragedy of commons. Suppose that individuals graze their flocks of sheep in a communally owned pasture. Individual sheep owners will be tempted to increase pasture. Individual sheep owners will be tempted to increase the size of their own flocks, since by adding a sheep or two, they will get more lambs and wool. If the optimum number is already in the pasture, their action will reduce only slightly the total yield and they will be ahead. But if all the other sheep owners behave selfishly and add a few sheep, the common pasture will soon be destroyed and all individuals will lose. The solution is for all the owners to come to a common agreement or convention, to limit the size of their flocks and forgo any short-term gain in favour of the long-term benefit to the whole community.

Behavioural Mechanisms

Epideictic Displays

Wynne-Edwards argued that the aerial maneuvers performed flocks of blackbirds or starlings in the evening are unrelated to feeding or defense against predators. He claimed that the main purpose of such social behaviour is to communicate information about the supply of potentially limiting resources. The behaviours, called *epideictic displays*, act as conventions resulting in adjustments of birth, death, and movement rates, enabling populations to track closely and not overexploit the fluctuating external environment.

Wynne-Edwards included among specially timed communal displays the dancing of gnats and midges, the maneuvers of bats and birds at roosting time, and the choruses of birds, bats, frogs, fish, insects, and shrimps. He believed that displays are abstract conventions, highly symbolic, and produce a state of "excitation and tension closely reflecting the size and impressiveness of the display." A special time of day might be set aside for them, as well as a special place. The mass feeding and roosting flights of birds are among the phenomena to which *Wynne-Edwards* applied his theory. Each night in the fall and winter, starlings (*Sturnus vulgaris*) gather by the millions. Birds in a roost use a common feeding area, and *Wynne-Edwards* considered it to be a communal territory.

Most birds are constant in their choice of roost and years. Each birds tends to have its own place in the roost. As the birds go to roost in the evening there is much communal singing and bickering over perching sets. As *Wynne-Edwards* noted: Moreover on fine evenings, especially early in the autumn, either a part or sometimes the whole of the noisy company not infrequently rises with a great roar of wings to engage in the most impressive aerial manoeuvres over the site: the massed flock may extend in a tight formation sometimes hundreds of yards in length, changing shape and direction like a giant amoeba silhouetted against the sky.

On the grand scale these manoeuvres are by no means the least of the marvels of animal adaptation, so perfect is their co-ordination and so intense the urge to excel in their performance. It seems quite irrational to dismiss what is certainly the starling's most striking social accomplishment merely as a recreation devoid of purpose or survival value, and wiser to assume that a communal

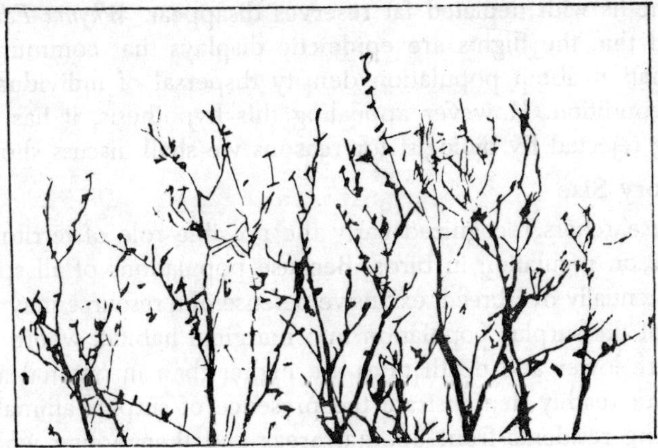

Fig. 10.3. Aggregation of starlings.

exercise so highly perfected is fulfilling an important function. Wynne-Edwards concluded that the primary function of the roost is to bring members of a population unit together for an epideictic display.

The result is to stimulate the adjustment of population density through either immigration of individuals of poor quality or a reduction in the reproductive rate. The roost's function as a sleeping place is secondary to its function as the location of the morning and evening displays. In fact, the concentration in the roost makes the birds an easier target for predators than they would be if they slept individually. Such synchronous mass behaviour is not limited to colonial birds.

The *evening rise*, characteristic of brown trout (*Salmo trutta*) among others, occurs about twenty to thirty minutes after sundown and lasts about thirty minutes. The fish cruise in small circles in a particular area. Although the rise is influenced by weather and the abundance of flies (food), the relationship is not very close, and Wynne-Edwards considered the rise to be epideictic. The Bogong moth (*Agrotis infusa*) of Australia forms large aggregations in caves in the summer. The moths become active for a half hour or so in the morning and evening. After vibrating their wings and crawling about, the moths fly around the cave and join others outside in a dense milling flight before they return to the same cave.

There is no evidence of any feeding or mating associated with this behaviour. The size of the population varies during the summer, and moths with depleted fat reserves disappear. *Whynne-Edwards* argued that the flights are epideictic displays that communicate information about population density dispersal of individuals in poor condition. However appealing this hypothesis, it has been largely rejected by biologist for reasons we shall discuss shortly.

Territory Size

Researchers recognized early the possible role of territory in population regulation in birds. Because populations of all species can potentially overbreed, exclusive defense of a resource may force some of the surplus population into marginal habitat, where birth rates are lower and death rates are higher than in optimal areas. One can readily demonstrate the presence of surplus animals by removing residents from optimal areas and then noting whether new animals come into the area from the marginal habitats. In his analysis of more than twenty years of breeding data for the great tit (*Parus major*) from Marley Wood in England, *Krebs* (1970) from population is regulated by density-dependent decreases in both clutch size and hatching success. He studied the role of territoriality in limiting the number of breeding birds as follows.

First *Krebs* (1971) demonstrated that defense of nest boxes causes the birds to be spaced out more than would be expected if nest box occupation were random, so that breeding density is reduced locally. Next he studied the effect of removing birds. The arrival of new pairs within hours after he had removed pairs of birds suggested that surplus birds were in the area. But where did these

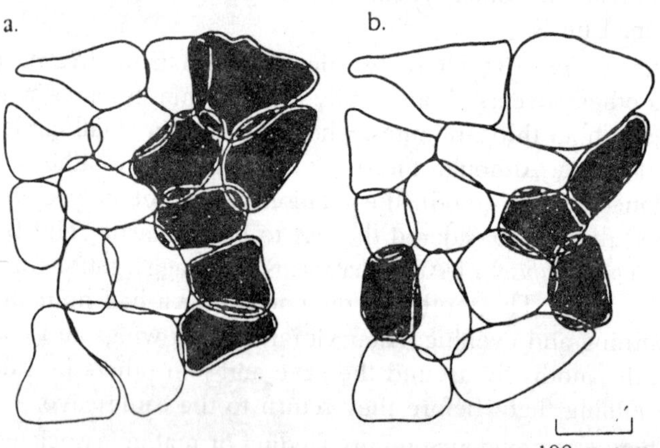

Fig. 10.4. Territorial replacement of great tits.

birds come from? Having colour banded most of the pairs in the area, Krebs found that nearly all the newcomers were from the hedgerows surrounding the woods.

The lower reproductive success of birds that breed in hedgerows compared with those that breed in the woods demonstrates again that territories limit breeding in the optimal habitat. However, Krebs found that although territory size varied considerably from year to year, it was not related to food supply. For this reason he concluded that territory is not very important in regulating the population as a whole, having only local effects. Several other lines of evidence suggest that territory size is related on limited resources and could therefore play a role in population regulation. The area defended by some species contains the minimum amount of resource to reproduce.

For instance, in a study on red squirrels (*Tamiasciurus* spp.), Smith (1986) measured the food supply in each territory and concluded that each territory contained just enough food to provide the energy needs of its occupant. Studies of a number of species have shown that territories are larger in marginal habitats. Nevertheless, to prove that territory regulates population is difficult. In polygynous species such as redwings male territory size is inversely related to habitat quality, but only other males are excluded and some feeding takes place off the territory. If the sex ratio is 50:50 and all females breed, it is not likely that male territory regulates the population as a whole. In white-footed mice (*Peromyscus leucopus*), male home ranges (which overlap to some extent and are therefore not strictly territories) decrease in size linearly with increasing density of males.

It is therefore not likely that home ranges regulate population. Female home ranges are smaller, overlap little with those of other females, and are inelastic above a certain density; thus they possibly set an upper limit to the number of nesting females. The prevailing view of the function of territoriality is that it benefits individuals, any role in population regulation is likely to be localized. There is little experimental support for the idea that territoriality regulates populations or that it has involved as a means of population regulation perse.

Dominance Hierarchies

Dominance hierarchies have also been implicated in the regulation of population. Lower-ranking individuals in groups with dominance hierarchies may have higher mortality and lower birth rates, as do surplus animals, those excluded from territories in prime habitats. In times of food shortage, high-ranking individuals survive while low-ranking individuals starve or disperse. Most group-living animals are polygynous, and the effects of competition on reproductive success are severe in the male. The elephant seal (*Mirounga angustirostris*) mating system demonstrates this effect spectacularly: the top-ranking male does as much as 80 per cent of the mating. However, a substantial turnover of males in the hierarchy takes place during the breeding season, so others do get a chance. Among redwings, nearly all females breed successfully each season; thus, the male-based dominance system is ineffective in population regulation.

Furthermore, in many other polygynous species with dominance hierarchies, the correlation between male rank and reproductive success is low. Sometimes the reproductive differential occurs in the female hierarchy. In temperate North America, paper wasp (*Polistes fuscatus*) nests are usually started each spring by a single queen, but she is joined by several other females. The foundress is dominant, and she suppresses reproduction in the subordinates, whose role is to regurgitate food for the foundress (trophallaxis) and to care for the young. In packs of wolves (*Canis lupus*) the alpha female inhibits mating by lower-ranking females; she is usually the only female in the pack to mate.

Drickamer and *Sade* and others found that among the rhesus monkeys (*Macaca mulatta*) on islands off the coast of Puerto Rico, matriarchies headed by high-ranking females are reproductively more successful than those headed by lower-ranking females. In fact, some of the latter actually lost numbers over the years. Although the preceding examples demonstrate density dependent reproductive restraint, they do not prove that dominance hierarchies function to regulate populations.

Social Pathology of Overpopulation

One research strategy for studying population regulation has been to create artificial populations in the laboratory. Such populations invariably cease to grow after following a roughly sigmoid (S-shaped) growth curve, even when food, water, and nesting space are kept in excess. *Calhoun* described the behavioural changes that take place in his mouse or rat *universes*. Initially, males defend territories, and birth and juvenile survival rates are high. As density increases, aberration begin. One is called a *behavioural sink*: animals become conditioned to eating and drinking in the presence of others and thus restrict their activities to a few places in the cage. In mouse universes, large numbers of *grouped withdrawn* spend nearly all their time in piles.

Occasionally fights break out, and the jumping mice resemble popcorn. Those females that bear young build inadequate nests and may abandon litters. *Calhoun* observed aberrant kinds of behaviour in male rats and mice in his *universes*, as well as the territorial, aggressive behaviour found in most natural populations. Pansexual males fail to discriminate between other males and females and make sexual advances to any creature that moves. Some males, called probers, are very active; although they do not compete for social rank or physical space, they show hypersexual, homosexual, and cannibalistic behaviour. Normally an estrous female rat, after being pursued by a male, returns to her burrow while the male waits outside. After the male performs a courtship dance, the female emerges and they mate. The probers, however, follow the female inside the burrow and sometimes consume dead young that are there.

Calhoun has refereed to another group of males as *beautiful ones*. These males have flawless coats with no scars and walk about the pen with impunity, being ignored by territorial males and even nesting females. At first, *Calhoun* thought these males were the highest ranking of all, but he later concluded conspecifics do not recognize them as adult rats or mice because they are in a state of physical and social immaturity even though of adult size. These males, when put in an uncrowded cage with receptive females, are unable to form dominance hierarchies, defend territories, or mate with females. Females in these colonies show delayed maturation, and few breed at all.

In fact, one of Calhoun's mouse universes went extinct because the females all became too old to reproduce without ever coming into estrus. Reproductive inhibition in response to density has also been demonstrated in confined populations of other species, such as deermice. Much attention has been paid to the role of agonistic behaviour and interaction rates in laboratory population of mice. Populations with particularly aggressive individuals tend to level off at lower densities than do those with less-aggressive members. When *Vessey* with the use of tranquilizers, lowered agonistic

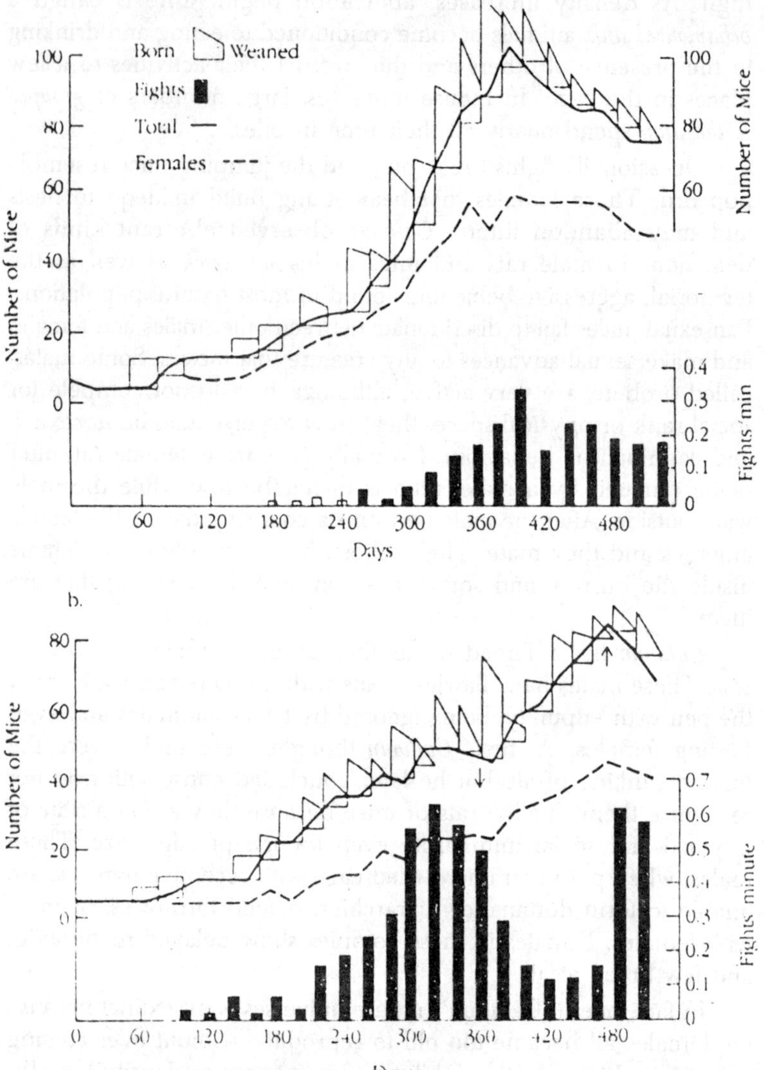

Fig. 10.5. Effects of tranquilizers on fighting, litter survival, and population growth of caged wild house mice.

behaviour of mice in populations that had reached the carrying capacity of the cage, litter survival improved and the population increased further. These experiments suffer from several problems that make extrapolation to natural populations risky: providing excess food, water, and nesting immigration which results in abnormally high densities. Some researches have tried to apply these findings to human populations.

A number of studies have shown positive correlations between density and such variables as crime rate and mental illness, but these correlations tend to disappear when the level of income is controlled. People living in crowded, urban areas tend to have lower incomes than do those who live in low-density areas, and it is this poverty, rather than density itself, that is casually related to crime and mental illness It also seems clear that sheer numbers per unit space are less important in social pathology than are the way space is utilized and the nature of the social interactions that take place. Although humans respond physiologically to stress like many other animals, there is not firm evidence that birth rates are reduced in response to stress. However, a number of reproductive restraints are evident in hunter-gatherer populations. Delayed onset of puberty, prolonged lactation (which delays resumption of ovulatory cycles), and preferential female infanticide are the main ways family size is kept in check. Presumably these factors work in conjunction with mortality and immigration to adjust populations to the carrying capacity. Lee (1980) emphasized the importance of prolonged lactation; and *Dickemann* (1975) has argued the importance of infanticide.

Behavioural Physiological Mechanisms

Some animals species undergo seemingly regular nine-to-ten-year population cycles as in the case of the lynx (*Lynx canadensis*) and the snowshoe hare (*Lepus americanus*), or three-to-four-year cycles as with the vole (*Microtus* spp.) and the lemming (*Dicrostonyx* spp. and *Lemmus* spp.). For example, during these population peaks a disturbance as slight as a hand clap was enough to send them into convulsions and coma, followed by death due to hypoglycemic shock. In 1950 Christian proposed that mammalian populations could be regulate by disease caused by exhaustion of the adrenal gland, following prolonged psychological stress from agonistic interactions at high population levels. This negative feedback loop, dependent on intraspecific competition, would be perfectly density dependent on intraspecific competition, would be perfectly density dependent.

The idea grew from the work of *Selye* on the *general adaptation syndrome*, in which nonspecific stressors, such as heat, cold, or defeat in a fight, produce a specific physiological response. ACTH released from the anterior pituitary, under control of the hypothalamus, stimulates production of glucocorticoids by the adrenal gland. These hormones, such as cortisone, function mainly to elevate blood glucose to prepare the body for fight or flight. The phenomenon of death due to adrenal exhaustion turned out to be an extreme case, and Christian modified his hypothesis after a series of laboratory experiments.

There is, in fact, an increase in adrenocortical output in response to increasing population density. These hormones, along with ACTH and some others still being explored, seem to reduce direct fitness in numerous ways. The body's two main defence mechanisms, the immune and the inflammatory responses, are inhibited; these changes obviously increase the likelihood of morbidity or mortality. At the same time, growth and sexual maturation are inhibited, as are spermatogenesis, ovulation, and lactation. Some of these effects on reproduction persist even into subsequent generations, in spite of a reduction in population density. Field data supporting these findings have come from studies of a variety of mammals, such as deer, rats, mice and woodchucks.

Probably an equal number of studies have failed to find a consistent relationship between adrenocortical output and population density. Part of the problem is that the stressor is not density, as seen earlier, but levels of agonistic behaviour in competition for limiting resources. In the laboratory, as little as two minutes per day exposure to trained fighter mice produces a pronounced stress response as well as fifteen fold increase in the load of parasites in experimental mice compared with control mice. Clearly, this negative feedback loop works under some conditions, but its generality in natural systems remains to be demonstrated. Work with pheromones has augmented some of the above findings. Substances released in the urine produce effects on conspecific without a physical encounter.

In some species of mice, the smell of strange male urine causes a pregnancy block by preventing implantation in crowded or unstable populations, the birth rate could thus be lowered. Mice avoid the urine of a mouse recently defeated by another mouse, and urine from stressed mice produces an adrenocortical response in naive mice. Females grouped together produce a substance that delays sexual maturation in other females the same delay has been

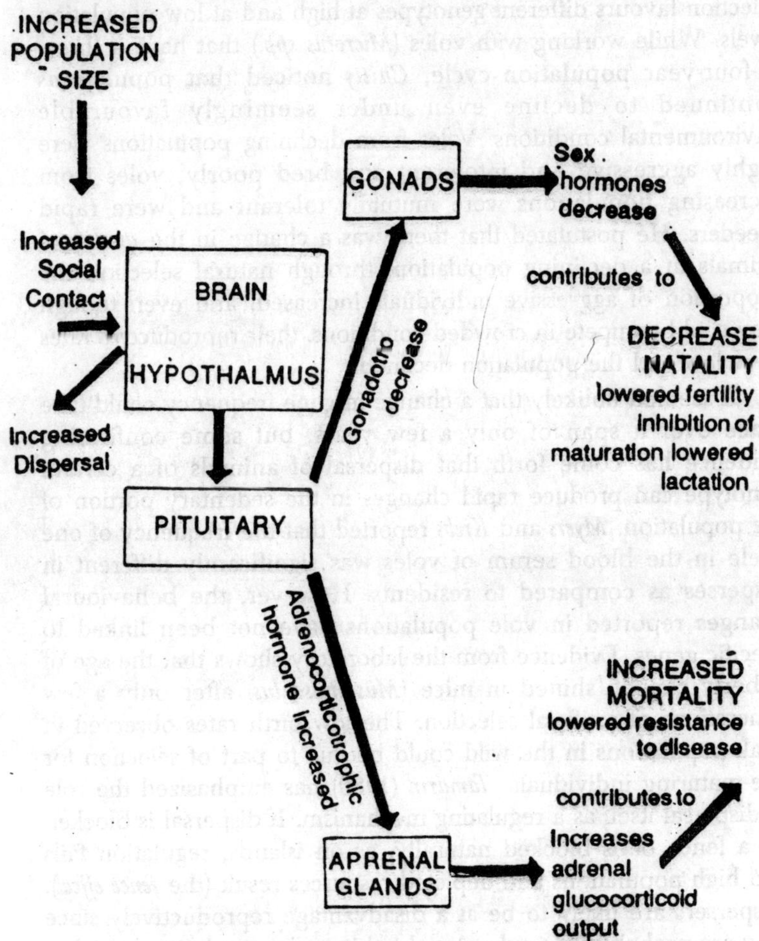

Fig. 10.6. Christian's model of population regulation in small mammals.

produced by using female urine from high-density populations of house mice in the field. In contrast, odour from a mature male accelerates the sexual maturation of young females. No single mechanism is likely to account for regulation of numbers in all species, and several mechanisms may operate within the same species.

Behavioural-Genetic Mechanisms

In general, the ideas examined so far are complementary; they are all based on the notion that increased interaction rates act to reduce population size. A competing hypothesis suggests that natural selection favours different genotypes at high and at low population levels. While working with voles (*Microtus spp.*) that have a three-to-four-year population cycle, *Chitty* noticed that populations continued to decline even under seemingly favourable environmental conditions. Voles from declining populations were highly aggressive and intolerant and bred poorly; voles from increasing populations were mutually tolerant and were rapid breeders. He postulated that there was a change in the *quality* of animals in a declining population; through natural selection the proportion of aggressive individuals increased, and even though they could compete in crowded conditions, their reproductive rates were low and the population declined.

It seemed unlikely that a change in gene frequency could take place over a span of only a few years, but some confirming evidence has come forth that dispersal of animals of a certain genotype can produce rapid changes in the sedentary portion of the population. *Myers* and *Krebs* reported that the frequency of one allele in the blood serum of voles was significantly different in disperses as compared to residents. However, the behavioural changes reported in vole populations have not been linked to specific genes. Evidence from the laboratory shows that the age of puberty can be shifted in mice (*Mus musculus*) after only a few generations of artificial selection. The low birth rates observed in peak populations in the wild could be due to part of selection for late-maturing individuals. *Tamarin* (1980) has emphasized the role of dispersal itself as a regulating mechanism. If dispersal is blocked by a fence or is blocked naturally, as on islands, regulation fails and high populations and depleted resources result (the *fence effect*). Dispersers are likely to be at a disadvantage reproductively since they are probably in a suboptimal habitat; also predation is higher on transient individuals. However, emigrating individuals have the potential of colonizing new areas and may be the founders of new species.

EVOLUTION OF POPULATION SELF-REGULATION

Proponents of self-regulation have not yet resolved two important problems. First because most of the data are from laboratory populations, where densities are often unrealistically high and were immigration is usually prevented, more testing on natural populations is needed. However, in natural populations many extrinsic factors come into play, and no single mechanism can be

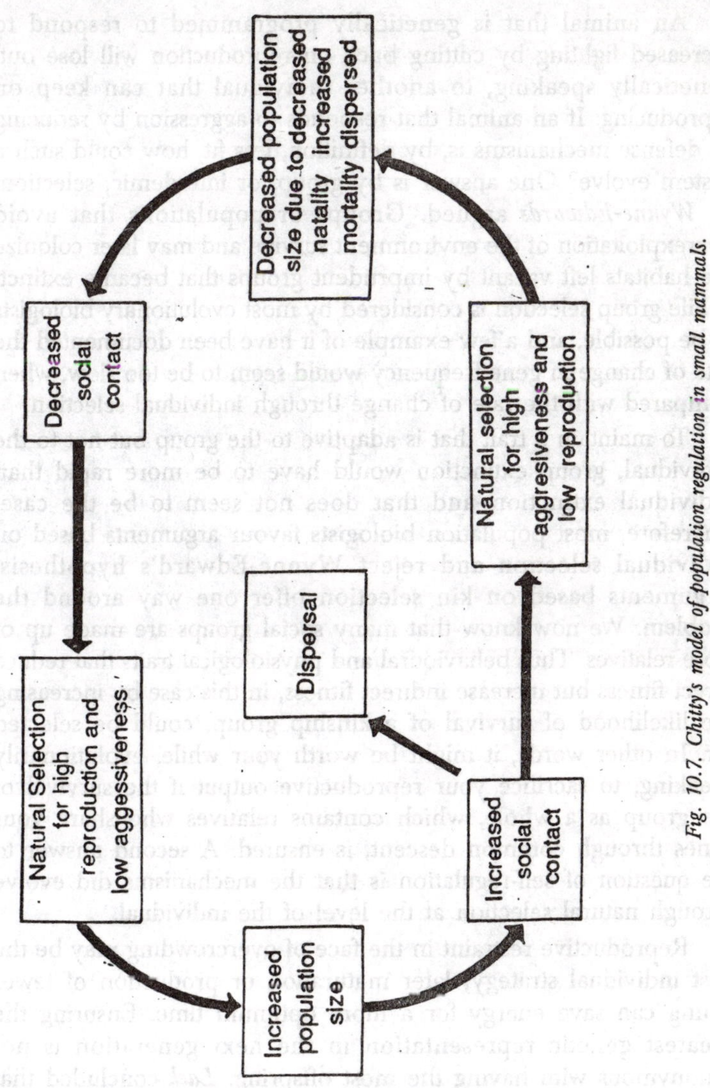

Fig. 10.7. Chitty's model of population regulation in small mammals.

implicated. Second, if we accept the idea that such self-regulatory mechanisms exist, do they involve specifically for the purpose, and, if so, what is the unit of selection? We can probably agree that self-regulation is adaptive for a population or species that has the potential for destroying or using up its resources, but many of the mechanisms we have discussed seem to be maladaptive for the individuals involved.

An animal that is genetically programmed to respond to increased fighting by cutting back on reproduction will lose out, genetically speaking, to another individual that can keep on reproducing. If an animal that responds to aggression by reducing its defense mechanisms is, by definition, less fit, how could such a system evolve? One answer is by group, or interdemic, selection, as *Wynne-Edwards* argued. Groups or populations that avoid overexploitation of the environment survive and may later colonize the habitats left vacant by imprudent groups that became extinct. While group selection is considered by most evolutionary biologists to be possible, and a few example of it have been documented the rate of change in gene frequency would seem to be too slow, when compared with the rate of change through individual selection.

To maintain a trait that is adaptive to the group but not to the individual, group extinction would have to be more rapid than individual extinction–and that does not seem to be the case. Therefore, most population biologists favour arguments based on individual selection and reject Wynne-Edward's hypothesis. Arguments based on kin selection offer one way around the problem. We now know that many social groups are made up of close relatives. Thus behavioural and physiological traits that reduce direct fitness but increase indirect fitness, in this case by increasing the likelihood of survival of a kinship group, could be selected for. In other words, it might be worth your while, evolutionarily speaking, to sacrifice your reproductive output if the survival of the group as a whole, which contains relatives who share your genes through common descent, is ensured. A second answer to the question of self-regulation is that the mechanisms did evolve through natural selection at the level of the individual.

Reproductive restraint in the face of overcrowding may be the best individual strategy; later maturation or production of fewer young can save energy for a more optimum time. Ensuring the greatest genetic representation in the next generation is not synonymous with having the most offspring. *Lack* concluded that clutch size (the number of eggs laid) in birds evolved to an optimum number in terms of the number of offspring that will, in turn, survive and reproduce; laying too many eggs could result in survival of few or no young. We can understand many of the epideictic displays described by *Wynne-Edwards* simply as outcomes of intraspecific competition of individuals that are behaving so as to increase direct fitness. Similarly, at times of high density,

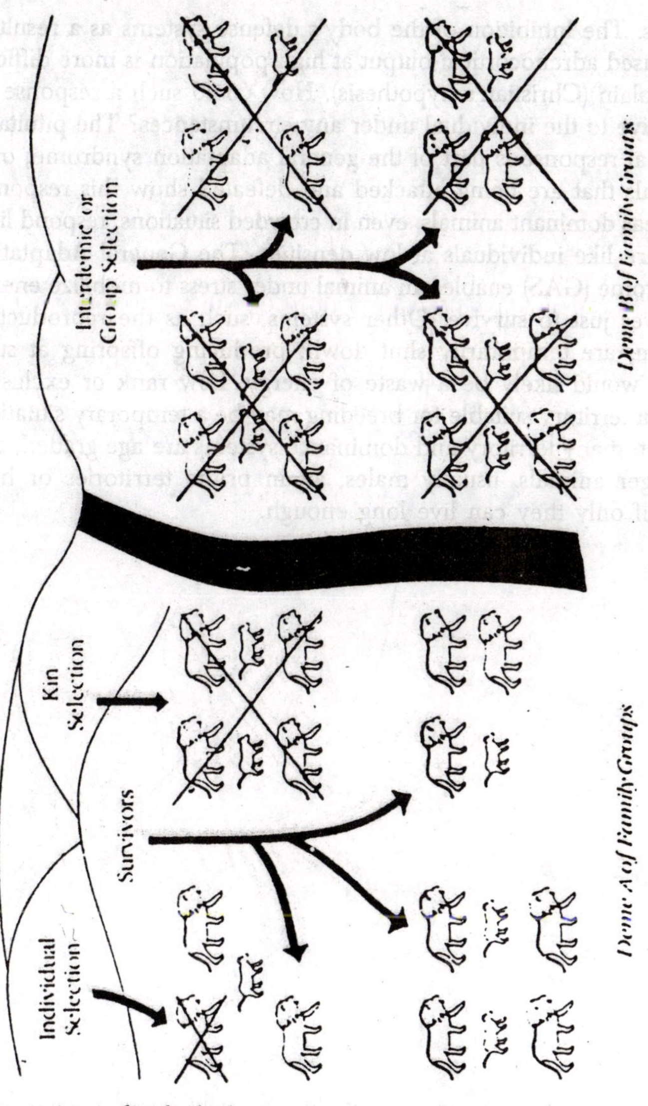

Fig. 10.8. Levels of selection, individual to interdemic.

aggressive individuals that are intolerant of others would be selected for it limiting factors were operating (Chitty's hypothesis). The most successful individuals at times of high density might be those that reduce the fitness of others through competition.

We should keep in mind that fitness is a relative term that refers to increasing one's genes in the next generation relative to others. The inhibition of the body's defense systems as a result of increased adrenocortical output at high population is more difficult to explain (Christian's hypothesis). How could such a response be adaptive to the individual under any circumstances? The pituitary-adrenal response is part of the general adaptation syndrome; only animals that are being attacked and defeated show this response, whereas dominant animals, even in crowded situations, respond little and are like individuals at low densities. The General Adaptation Syndrome (GAS) enables an animal under stress to mobilize energy reserves just to survive. Other systems, such as the reproductive system, are temporarily shut down; producing offspring at such times would likely be a waste of energy. Low rank or exclusion from a territory suitable for breeding may be a temporary situation. In fact, many territory and dominance systems are age graded, and younger animals, usually males, attain prime territories or high rank if only they can live long enough.